Lecture Notes on Data Engineering and Communications Technologies

Volume 119

Series Editor

Fatos Xhafa, Technical University of Catalonia, Barcelona, Spain

The aim of the book series is to present cutting edge engineering approaches to data technologies and communications. It will publish latest advances on the engineering task of building and deploying distributed, scalable and reliable data infrastructures and communication systems.

The series will have a prominent applied focus on data technologies and communications with aim to promote the bridging from fundamental research on data science and networking to data engineering and communications that lead to industry products, business knowledge and standardisation.

Indexed by SCOPUS, INSPEC, EI Compendex.

All books published in the series are submitted for consideration in Web of Science.

More information about this series at https://link.springer.com/bookseries/15362

Zhengbing Hu · Bo Wang ·
Sergey Petoukhov · Matthew He
Editors

Advances in Artificial Systems for Power Engineering II

Springer

Editors
Zhengbing Hu
Faculty of Applied Mathematics
National Technical University of Ukraine
Kyiv, Ukraine

Bo Wang
School of Electrical and Automation
Wuhan University
Wuhan, China

Sergey Petoukhov
Mechanical Engineering Research Institute
Russian Academy of Sciences
Moscow, Russia

Matthew He
Halmos College of Arts and Sciences
Nova Southeastern University
Florida, FL, USA

ISSN 2367-4512 ISSN 2367-4520 (electronic)
Lecture Notes on Data Engineering and Communications Technologies
ISBN 978-3-030-97063-5 ISBN 978-3-030-97064-2 (eBook)
https://doi.org/10.1007/978-3-030-97064-2

This Springer imprint is published by the registered company Springer Nature Switzerland AG
The registered company address is: Gewerbestrasse 11, 6330 Cham, Switzerland

Preface

The development of artificial intelligence systems and their applications in various fields belongs to the most urgent tasks of modern science and technology. One of these areas is energy and power engineering, in which their application is aimed at increasing the effectiveness of generation and distribution of energy and power for the life support of the world's population, including the tasks of developing industry, agriculture, medicine, transport, etc. The rapid development of artificial intelligence systems requires the intensification of training of a growing number of relevant specialists. At the same time, artificial intelligence systems have significant perspectives of their application inside education technologies themselves for improving the quality of training of specialists taking into account personal characteristics of such specialists and also the emergence of new computer devices.

In digital systems of artificial intelligence, scientists try to reproduce the inherited intellectual abilities of humans and other biological organisms. The profound study of genetic systems and inherited biological processes can reveal bio-information patents of living nature and give new approaches to create more and more effective methods of artificial intelligence. For this reason, intensive development of bio-mathematical studies of genetic and sensory-motor systems is required on the basis of contemporary achievements in mathematics, computer and quantum informatics, biology, physics, etc. In other words, the study of genetic systems and the creation of methods of artificial intelligence should go hand in a parallel manner to enrich each other.

The Second International Conference of Artificial Intelligence and Power Engineering (December 17–19, 2021, Moscow, Russia) has the purpose of presenting new thematic approaches, methods, and achievements of mathematicians, biologists, physicians, and technologists and also attracting additional interest from different specialists in this prospective theme. Its proceedings also include articles on specific tasks in various fields where artificial intelligence systems can be applied in the future with great benefit.

The conference is organized jointly by the Mechanical Engineering Research Institute of the Russian Academy of Sciences (IMASH RAN), the "International Research Association of Modern Education and Computer Science" (RAMECS), and Wuhan University with the participation of the Polish Operational and Systems Society.

The organization of such conferences is one example of growing Russian–Chinese cooperation in different fields of science and education.

The best contributions to the conference were selected by the program committee for inclusion in this book out of all submissions.

We are grateful to Springer-Verlag and Fatos Xhafa as the editor responsible for the series "Lecture Notes on Data Engineering and Communications Technologies" for their great support in publishing the conference proceedings.

Zhengbing Hu
Sergey V. Petoukhov
Matthew He

Organization

Conference Organizers and Supporters

Wuhan University, China
Mechanical Engineering Research Institute of the Russian Academy of Sciences, Russia
International Research Association of Modern Education and Computer Science, Hong Kong
Qinghai University, China
International Center of Informatics and Computer Science, Ukraine
Polish Operational and Systems Society, Poland

Contents

Conceptual Model of Agro-Industrial Cyborg

Oleg N. Gurov(✉)

Institute of Industrial Management of the Russian Presidential Academy of National Economy and Public Administration, 82, bld. 3, Ave. Vernadsky, 119571 Moscow, Russian Federation
gurov-on@ranepa.ru

Abstract. The article explores agro-industrial complex in the context of technological development. The author makes an attempt to predict the radical introduction of digital technologies in agriculture, which will affect the status of the agricultural worker. The author introduces the concept of "agrocyborg"—a projected model of an agricultural worker who integrates into digital technologies. The upcoming transformation will make it possible to preserve traditional social characteristics and acquire new highly demanded qualities. Significance of this work is illustrated by an attempt of giving a philosophical and cultural understanding of the phenomenon of "agrocyborg" as a perspective conceptual structure, valuable for understanding the future of the development of the agro-industrial complex, in view of the development and implementation of digital technologies. As a result, the author comes to the conclusion that the concept of "agrocyborg" is not only the intersection of agricultural tradition and technological innovation, but also the starting point for the development of technologies in agriculture for the benefit of human development.

Keywords: Cyborg · Agrocyborg · Digital transformation · Agro-industrial complex · Digital transformation · Meta-parasite

1 Introduction

The concept of a cyborg has been actively elaborated in scientific and cultural discourse in recent decades. Cyborgization is described as the process of merging the corporeal with the technological, and combining by this the natural with the artificial. The purpose of this work is to conceptualize the "agrocyborg" phenomenon as a promising model of participation in the agro-industrial complex, which will combine the humanitarian values of a traditional countryman with technological capabilities. Such synergy allows for minimizing the risks of loss of humanity when introducing technologies into public life. At the same time, it adequately responds to the challenges which humanity is facing described in the UN SDGs. This article constitutes the part of the author's research on the boundaries of the human and human's prospects within the technogenic civilization [16]. The essential research methodology is based on analytical, systemic and synergistic approaches, together allowing revealing the depth of the context and providing for required focus on the ongoing interdisciplinary research.

The topic of cyborgization in scientific discourse has been already studied for quite a long time, starting from the moment when the concept of "cyborg" was introduced

Z. Hu et al. (Eds.): AIPE 2021, LNDECT 119, pp. 1–13, 2022.
https://doi.org/10.1007/978-3-030-97064-2_1

into scientific circulation and subsequently into mass culture by researchers M. Clynes and N. Kline [7].

Currently, this subject is widely represented in the framework of research on various aspects of technological development and human interaction with technology [26]. Current research refers to both fundamental and applied science. It is focused on actual social, economic, political and scientific aspects of technology implementation. In particular, some researchers pay attention to the negative aspects of the diffusion of technologies - for example, such as the prospects for robotization, bringing risks of unemployment for the service sector [19]. Other researchers consider the development of robotics and cyborgization as the process of physical improvement with the help of technological tools. Thereby it gains advantages over both the "traditional" man and the robot [4]. Promising research is being carried out in the field of on AI and neural networks, as core elements of the future social and economic system, and agro-industrial complex in particular [18]. Recent studies suggest robotization and cyborgization as a fait accompli. They are devoted to fairly pragmatic studies of how society reacts to cyborg and robot aesthetics in assessing the prospects for service interaction in certain business areas, which might affect people's perception and comfort while interacting with robots [25]. The problem of transformation of human into a cyborg is considered as a social process that is part of the development of society [11].

The breadth of the studied contexts confirms the importance of the research presented in this article. In this work, an attempt was made to implement a philosophical and culturological understanding of the concept of "agrocyborg", as well as impose it on the practical tools of technological improvement of the agro-industrial complex. Thereby it provides for the spiritual, ideological and cultural content of this important type of economic activity. This work provides basis for the formation of a continuity between traditional agricultural labor and promising agro-tech.

The novelty of the presented was aimed at finding a common denominator and common foundations for humanitarian ideas and practical activities, cultural and historical foundations and strategies for technological development in the field of the agro-industrial complex.

2 Cyborg as a Bridge Between Natural and Man-Made Origins

As mentioned above, cyborgization is the convergence of the human with technology, who possesses a new nature, and whose unity of consciousness and corporeality is complemented by technology, fully fitting into the brand-new model.

The grounds and motivation for cyborgization differ. On the one hand, it may be compensation for deficiencies or damage, overcoming the insolvency and dysfunction of the body, on another - striving for improvement, overcoming limitations and empowering. Given the progress of technologies and its spread in all spheres of social life, we do not allocate these grounds to opposite sides, considering no contradiction in different motivation. Turning to the origins of the "cyborg" concept, this invention of M. Clynes and N. Kline owes to their research of required conditions for humans during long-term space flights, and therefore of required characteristics to space explorers. It is important that scientists, describing this concept, focused on the fact that space exploration was primarily a spiritual challenge, and biological evolution was a prerequisite for human improvement in a high-stakes game [7].

This unusual angle of view is not innovative, it is essentially the eternal dispute between humanitarians and technocrats, urban and rural—such disputes during human history often determined the development of civilization, verifying the actual and falsifying the untenable categories of philosophy and culture.

For our study, the most relevant is the axiological aspect of cyborgization, based on value system standpoint, formed as a result of technosphere development, profoundly changing human culture and mentality. In these conditions the very mental content is being modified dramatically, and in this sense, the axiological direction is top prospect, since "… the" philosophical" problems of new forms of "spiritual life of man and society" are being solved, the explication of "moral universals" in the context of the "noosphere" is carried out, taking into account the "strategies of sustainable development of civilization`"" [2]. The COVID-19 pandemic makes this aspect especially acute, as struggle against pandemic changes the logic of social systems, while advanced technologies are used for the fight for lives of people and the well-being of the world [33].

Cyborgization affects various areas, which we will bring to two generalized groups for the sake of convenience:

- biological—in terms of replacing or supplementing human functions or organs with prostheses, implants, exoskeletons, navigators, etc.;
- intellectual—in terms of replacing or supplementing human intellectual abilities with smart machines.

It should be noted that some researchers stay extremely negative about such prospects. J. Baudrillard wrote that the human recognizes his powerlessness and incompetence by inventing complex mechanisms. "To entrust your intellect to a machine means to free yourself from any claim to knowledge" [3]. Developing this thesis, other scientists consider cyborgization as the disablement and autotomy of physical bodies in the process of human's technological expansion [9]. Noting the risks of dehumanization, A.Yu. Alekseev mentions that a number of researchers are afraid of dominance of artificial over natural, since it will prove that human was relevant only in the past, before the era of cyborgization, "… before technical programmable means began to join in our conscious activity". According to pessimists, humanity will turn into an army of zombies if chipization is not stopped [1].

However, we adhere to optimistic attitude towards human prospects, since we are close to J. Wales approach, who predicts that cyborgization will improve human abilities to adapt in the face of constantly changing and growing challenges [37]. In this sense, we subscribe to the opinion, which may sound somewhat speculative, that cyborgization began with the first tool of labor. The use of the most primitive tool in order to influence the world around us had an impact not so much on the strength and dexterity of human hands, but on the very mentality of a human and on his ability to interact with the world around in in its new capacity. Returning to the stated topic, the very first tool of labor, a biface (cleaver), was mainly used for foraging, which later transformed into agriculture [23]. The fourth industrial revolution (Industry 4.0) taking place today sometimes referred to in the media and in popular culture as the "revolution of robots", includes the "Human-Robot Collaboration" program. Under these conditions, a human is required to evolve or, in other words, to turn into a technogenic hybrid. This requirement is driven by the need to effectively integrate into the infosphere and interact with advanced technologies which increasingly define the production, economic and everyday landscape.

3 From the Plough Ad Astra and Back

50–70s of the XX century is a period that carried out the embryo of a cyborg in utero of a technological boom. During these years, humanity showed in practice that it was ready for anything to overcome its own borders, to go beyond, mastering and colonizing new worlds. However, in subsequent years, extensive development in this area did not follow, the human himself did not physically move beyond the Moon. The dreams subsided, the ideals of a consumer society triumphed in life, and space exploration began to be justified by considerations of economic expediency. The decline in interest in space exploration was accompanied by a decrease in funding for astronautics. The current generation is not a bearer of the ideas of space exploration. Space programs are being implemented slowly and not as a top priority. Current key projects are mostly associated with anchoring in near-earth orbit. It creates significant obstacles to further large-scale development of astronautics.

In the coming years, it will become obvious whether private initiatives will be able to fill the niche that we tend to assess not only as an ideological and technological field, but also as a space for the spiritual growth of humanity. Optimism is inspired by a retrospective look at the history of travel, exploration, wandering and expansion. Throughout history, great explorers and conquerors have sought to find not only new lands with their resources and treasures, but also new sources of knowledge and inspiration. It is possible that in the predictable future, a promising direction will be the study of exoplanets, some of which, being terrestrial planets, can become objects of colonization and, in particular, the development of the agro-industrial complex. However, for the current moment, humanity is faced with urgent tasks in the contours much narrower than space travel. An attempt to escape into outer space from the problems, which solution is required here on Earth, today and now, is fraught not only with the impossibility of further development of civilization, but even with the fatal consequences of the Anthropocene.

4 Sustainable Development Goals—The Circle is Completed on the Agro-Industrial Complex

Since the publication in 1987 of the report "Our Common Future", prepared by the UN Commission on Environment and Development, the concept of sustainable development, based on the idea of responsibility to future generations, has become the main postulate of achieving a prosperous future [31].

In 2015, the UN identified 17 sustainable development goals (SDG), formulating the main directions of work to solve the critical problems of civilization:

- Elimination of poverty;
- Preservation of the Earth's resources;
- Achievement of prosperity [33].

Such values of technogenic civilization as perception of nature as human activity object of transformation, and consumerist morality aimed at unbridled hedonism and triumph of consumption, have demonstrated their inconsistency and futility. Linear thinking, race to the top and technologic advance at any cost are replaced by the awareness of the need to transform thinking towards the world perception as an all-encompassing self-regulating meta-system that does not contradict the logic of cyclical development. Speaking about cycles, we cannot ignore the concept of agrarian cyclical time inherent in the medieval mentality with the opposition of "darkness and light, cold and warmth, life and death, activity and idleness" [15]. Today this approach is likely to acquire new relevance, primarily in the context of the dichotomy which defines a cyborg in similar oppositions. In addition, in a globalized world, most of the SDGs are, in one way or another, related to the need to improve the efficiency and boost development of the agro-industrial complex. Almost a third of the 17 SDGs are directly related to the development of the agro-industrial complex:

- Elimination of hunger;
- Poverty eradication—poverty manifests itself to a great extent in hunger and malnutrition;
- Good health and well-being—is largely conditioned by a healthy lifestyle, an important part of which is nutrition;
- Clean water and sanitation—2 clusters of problems can be identified for this goal: poor water quality jeopardizes food security, and drought exacerbates hunger and malnutrition;
- Responsible consumption and production—inefficient harvesting and transportation turns produced food into waste;
- Combating climate change—food and water shortages are largely conditioned by this problem [13]. Obviously, these goals directly determine the need to develop and optimize the agro-industrial complex [28].

5 Countryman Conservatism

Along with other spheres, the agro-industrial complex has recently been actively transforming under the influence of technology, but we are inclined to believe that the agricultural worker still retains special characteristics that are unique to representatives of this social type. L.V. Milov, exploring Russia and the Russian countryside in the context of the unity of nature and history, notes that historical agrarian Russia was a special society with a minimum amount of surplus product, in light of which constant hard work was vital for the survival of both the individual and the society as a whole. At the same time, the author notes the countryman' special way of life, which requires an almost constant exertion of strength and special endurance [27]. In a work devoted to the study of the characteristics of Russian countryman in the pre-revolutionary period, E.G. Sinyakina also distinguishes among the main qualities of the countryman mentality hard working nature and endurance, concentration on constant and hard work [32]. Creating a portrait of a countryman (farm labourer) of the 20th century, Canadian historian L. Viola examines the struggle of the Soviet peasantry against collectivization, and notes that the resilience and endurance of the peasantry are the main characteristics of countryman culture [36].

Thus, we see that for many centuries, despite numerous political, economic and technological transformations, the countryman remained conservative and committed to his characteristics. The study of the metaphysical and spiritual aspects of the agricultural sphere is outside the scope of this study. However, we must pay attention to the fact that the countryman's endurance and constant work at the limit of his capabilities was fueled by his connection with the "field", which is a sacred space between life and death, a place for testing spiritual and moral foundations – here we have one more dichotomy [34]. So, it makes sense that the parable "The Monk and the Countryman", written, apparently, in the 21st century, shows that it is the countryman who has a direct contact with God: "... The countryman was getting up early in the morning with the words: "good Lord" and was walking into the field. After plowing there all day, he was going to bed with the same words: "good Lord"... So, the countryman feeds himself and his family, and feeds all of us with his toil and trouble, and contacts God twice a day" [29].

To be fair, we note that Freud wrote that technological progress led to God-likeness of man: "... this is the direct fulfillment of all—no, of the majority—of fabulous wishes: man created all this by means of science and technology on earth, appearing on it like a weak animal at first" [12]. The modern researcher S. Midson pushes this statement to the limit, arguing that the appearance of a cyborg challenges the idea of man as a being made in God's likeness [21]. We will write further about the "nature" of the cyborg in the context of agricultural culture, and now we will hypothesize that cyborgs are different. For example, a cyborg used for military affairs may be fundamentally different from a cyborg for the agro-industrial complex, whose worldview will successfully fit into the paradigm of the traditional farm labourer.

6 Agro-Industrial Complex Transformation

The above considerations may lead to the false conclusion that the sphere of the agro-industrial complex is conservative and rigid, which is not true, since the agro-industrial complex is already a system of interaction of numerous biomach systems, encoded by the triad "man-machine-biotic" [6]. Among the main applied technologies used in the agro-industrial complex for the needs of animal breeding and crop farming, farm management, logistics and marketing, are:

- remote control using satellites, drones and sensors;
- tools for collecting and analyzing big data;
- robotization;
- internet of things [21].

As an illustrative example, we take the product of US Carbon Robotics company, which has recently released a series of farm robots whose function is to destroy weeds without harming the soil. One robot, which uses the thermal energy of a laser, is able to destroy 100,000 weeds in an hour. Such innovations undoubtedly have a serious effect on the activity of the agro-industrial complex right now.

Note that this is not the only example of practical application of robotics manipulators for precise weeding, pruning or harvesting. Moreover, the prospects for this market are colossal. According to the commercial study Future Growth by 2026, prepared in 2021, the global Agriculture Robot market size is projected to reach USD 4960.6 million by 2026, from USD 2349.1 million in 2021 [14].

According to the UN forecast, the world's population will reach 9.7 billion by 2050 [35]. In these conditions, increasing the efficiency of the agro-industrial complex through its qualitative modernization becomes a complex and strategic task which complies political, economic and social components. However, we are forced to argue that the change in the agricultural system in accordance with the logic of a smart factory based on the implementation of "… the Internet of things, deep learning neural networks, blockchain, big data analysis and other technologies does not qualitatively distinguish it from ordinary production and is not sufficient to provide an effective basis for a breakthrough digitalization of the economy, including the agro-industrial complex" [5]. Among these technologies, special attention is paid to the development of the neural network due to the prospects for development and application in the agro-industrial complex [24].

7 The Agrocyborg—The Farm Labourer of the Future

The above agroindustrial tendencies are fraught with the big risk of eliminating human, breaking his connection with the original roots—the planet, the land, the production of alimentary products. Realizing that progress cannot be stopped, and the cost of the success of agricultural system transformation is the survival of mankind, we assume that a compromise subject, an AGROCYBORG, will emerge on the "agricultural field". Agrocyborg will combine all the strong and promising characteristics of human and technological nature.

Based on existing definitions of cyborg in Oxford English Dictionary and other popular dictionaries of modern English language, we propose the following determination:

Agrocyborg is an agricultural worker (farm labourer) whose physical capabilities or abilities surpass the capacities of an ordinary human being by force of integration the regulatory devices and other technology products, broadening his vital functions.

The symbiosis between a persistent and hardy man of a traditionalist nature, being literally a cultural archetype, and an agrobot, which has already successfully proven itself in practice, is an interesting prospect, given that the common denominator will be the very endurance and hardworking nature that characterize both a farm laborer and a robot.

The new agricultural worker will be capable to apply new knowledge, skills and capabilities acquired by cyborgization in the following areas:

- constant connection and interaction with drones will allow monitoring the state of the farm, collecting data for subsequent analysis and prevention of emergencies;
- direct "participation" in the activities of software systems in terms of processing and analyzing data generated in the course of activities, within the framework of predictive analytics;
- deep integration of the agricultural worker into "precision farming", which results in a multiple increase in yield;
- becoming an agricultural worker as a full-fledged participant in the marketplace (platform, ecosystem) for the sale and delivery of products to consumers with exclusion of intermediaries, as well as for the sale and purchase of equipment and machinery [22].

We see that the proposed directions do not fundamentally change the philosophy of agriculture, but, on the contrary, "immerse" the agricultural worker in the traditional functionality to a greater extent, ensuring the continuity of activities that have developed over the centuries.

A separate promising area requiring special analysis is the integration of an employee into an automated work cycle of an agricultural enterprise, controlled by AI technologies [30]. This direction can be studied in the context of forecasting interaction between humans and AI, depending on the level and scale of intelligent systems utilization.

Going back to the original concept of cyborg in the context of space exploration, B. Flaherty writes that being the main constituent and integral part of the cyborg, man is the weak component, since he is not a machine: "Man is a sea-level, low-speed, one-g, 12-h animal" [10]. Despite that this statement can be perceived as discrimination against human, we believe that this is not about the prospects for dehumanization, but, on the contrary, about a qualitatively new round of human development. Let's reiterate that, according to M. Clynes and N. Kline, the cyborg personifies the complete freedom of human, and travel into space "… challenges mankind not only technologically but also spiritually, in that it invites man to take an active part in his own biological evolution" [7].

We also note the issue of preserving human autonomy in the technologized world. Robotization' growth dynamics in the agro-industrial sector is limited by high price of the technologies for most farmers, especially those who manage small farms, and currently the development is carried out in the logic of economics of scale. Under these conditions, agrocyborgization can be perceived as an investment in farmers' independence and in maintaining their autonomy, as a tool to protect against the need to enlarge farms and unite with other farmers.

In current context, agrocyborgization is not a process of introducing artificial elements into the human body that expand his capabilities, and connecting consciousness to information and analytical systems, but a new level of integration of human with nature, of a countryman with his land, a return to the unity of nature and freedom inherent in ancient culture.

Our optimism and sober attitude towards the possibilities of dehumanization is based on the forecast of further economic and social civilization development within the framework of the so-called Industry 5.0. The model being formed today is determined by focus on human-centeredness and sustainability. To address global challenges and ensure successful development, it is necessary to take into account the social and environmental aspects of development. In this paradigm, basic human interests and needs are the starting point for paths selection. The very approach to technology is changing—instead of requiring an employee to adapt his qualifications in accordance with the needs of technology, within Industry 5.0, technology itself adapts to human needs. The importance of investing in human capital, in improving human skills and abilities, is increasing [20].

What will the agrocyborg look like, how will he materialize in a practical field? We know that a significant part of the inventions defining today our everyday life are of military origin. These technologies include digital cameras and GPS, microwave ovens and, in fact, the Internet. Autonomous combat systems are being actively developed today. The use of such latest developments as armed drones has already become a problem at the level of international relations, not to mention an ethical dilemma which is far from being resolved. Cyborgization of military personnel is being developed as well. In 2019, the U.S. The Army's Combat Capabilities Development Command has published a study with plans for the coming decades to modify military personnel to perform service and combat missions [8].

We will use several promising military technologies to extrapolate it to the agrocyborg model:

- Introduction of optical systems which allow visualizing the territory in a variety of wave ranges and identifying various specified targets;
- Increase in physical strength and endurance due to the introduction of "optogenetic musculoskeletal control system";
- Improving the human brain through the introduction of cybernetic systems which will provide remote control of machinery and equipment.

8 Agrocyborg as an Effective Fighter Against Meta-parasitic Phenomena Immanent in Modern Civilization

The relevance of agrocyborg model is mandated by not only his high-quality working abilities, but also by the fact that agrocyborg is able to resist infodemic tendencies, meaning that information epidemic spreads in the world as quickly as the biological one, and the consequences of both of them are critical for civilization welfare. Fake news, manifestation of hatred, radical political phenomena, panic and digital fatigue are the problems that largely determine current cultural landscape. Infodemic tendencies are caused by the meta-parasitic logic of the spread of information flows in the world, where technologies play the key role, along with that the level of digital literacy significantly lags behind the required one.

Meta-parasite appears as a result of synthesis of natural, biological and socio-cultural phobias, fears and panic phenomena, which have acquired an all-encompassing scale and begin to live their own lives. This concept has become possible as a result of sharp and rapid improvement and spread of digital technologies, due to which the key phenomena and tools of modern life overcame their natural boundaries, and from political, social, agricultural and medical spheres got intertwined into a single tangle. COVID-19 pandemic has demonstrated that meta-parasitic concept can be applied to characterize the situation around the world. Today we are witnessing formation of a real meta-parasitic construct on a global scale—there is a virus, an infection in the physical world. And in parallel with it, the virus appears in informational and cultural space. These toxic and virulent phenomena are turning into fertile ground for destructive trends in international relations, economic processes and culture. Thus, the meta-parasite is the result of the diffusion of various (biological, informational and social) hypostases of the terrifying parasite, which multiply and acquire new dimensions within the viral system of information and communication technologies.

The general concept of parasitism is largely related to the agricultural sector, with the need to constantly fight and protect the economy from parasites. The traditional agricultural worker is by nature an effective "pest control agent" in his field on a daily basis. By the same logic, an agrocyborg can become an equally effective neutralizer of meta-parasitic phenomena, being one of the main risks for the development of civilization. Thus, the significance of the agrocyborg goes beyond the economic discourse to reach ideological and civilizational scales. In this capacity, the agrocyborg will be able to contribute to overcome the large-scale meta-parasitic construct that destroys foundations of humanism, which is the main factor in the successful development of further human civilization.

9 Conclusions

D. Haraway defines cyborg as a creature belonging to the cosmic postgendered world, who lacks origins, attachments or addictions [17]. The proposed concept of agrocyborg has a different dimension. All global innovations enter public life not instantly, but within certain period of time, gradually getting "assimilated" and "humanized" during this time. In the context of agriculture, this principle is most indicative—the spread of cyborgization is supposed to change the agro-industrial sector, while the agrocyborg will integrate the whole range of human properties.

As an example, it perfectly reflects in mass culture, convincingly testifying that these issues are in demand in public life. In the fall of 2020, the Russian short film "Russian Cyberpunk Farm" by S. Vasiliev was posted online. The plot of the short film combines the realities of a traditional Russian village with sci-fi technologies of the future against the background of cybercows and cyber-maidens backed by cyber-folk songs. The video went viral - nearly 10 million views (as of May 2021), The wave of rave reviews indicated that the idea of combining the latest technology with the preservation of the traditional rural lifestyle in the village does not cause cognitive dissonance and responds positively in the minds of modern people.

From history it is known that sci-fi and humorous ideas often come true, thus the agrocyborg concept seems to be viable and promising.

As a result of this study, it is shown that the agrocyborg concept has certain philosophical and cultural grounds and could be used as bases for further research and forecasting on the development of the main technological areas of the agro-industrial complex. The agrocyborg serves as a model for future that combines the achievements of human culture and civilization allowing for preserving the values accumulated in the course of civilization development. At the same time agrocyborg generates new opportunities arising from the technological progress, in the most vital area for survival and development of human civilization – the agro-industrial complex.

Theoretical viewpoint tells us, that the results obtained by the study could be used as a starting point for further discussion on the interaction between a human and technology, organic and artificial (digital and mechanical). For practical purposes, the obtained results should be taken into account by specialists in the agro-industrial complex while assessing the social consequences of advanced technologies integration into the basic structure of the agro-industrial complex. Along with this, the agrocyborg concept can be also useful for technical specialists developing applied tools for the agro-industrial complex, such as agrobots, precision farming and precision livestock systems, neural networks and others.

The work done and the results obtained require more in-depth study in terms of quantitative indicators related to specific technologies serving as the basis for agro-cyborg in the future. It seems to be also promising to expand on the interdisciplinary space for studying this problem, including sociological vision for assessment by social science.

Acknowledgements. The author sincerely thanks Dr. phil., Professor A.Yu. Alekseev for invaluable help in scientific research and for fruitful and inspiring discussion, resulted in the birth of the presented agrocyborg concept.

References

1. Alekseev, A.Yu.: "The third way" to electronic culture. Filosofskiye nauki (Philos. Sci.) (2), 65–71 (2017)
2. Alekseev, A.Yu.: Multiagent categories of roboPhilosophy: the roboPerson and a roboZombie. Artif. Soc. **15**(2) (2020). https://doi.org/10.18254/S207751800009761-6
3. Baudrillard, J.: Transparency of evil. Dobrosvet, 400 p. (2006)
4. Carvalko, J.: The Techno-Human Shell: A Jump in the Evolutionary Gap. Sunbury Press, Mechanicsburg (2012)
5. Chernoivanov, V.I., Tolokonnikov, G.K.: The concept of "smart enterprise" in the paradigm of biomass systems. Tekhnika i oborudovaniye dlya sela (Rural Mach. Equip.) **10**, 2–7 (2018)
6. Chernoivanov, V.I., Tolokonnikov, G.K., Melnikov, V.A.: Digitalization of the agro-industrial complex in the paradigm of biomash systems. Bull. VNIIMZh (2), pp. 27–33
7. Clynes, M.E., Kline, N.S.: Cyborgs and Space. Astronautics, pp. 27–31, September 1960

8. Cyborg warriors could be here by 2050. DoD study group says. https://www.armytimes.com/news/your-army/2019/11/27/cyborg-warriors-could-be-here-by-2050-dod-study-group-says/. Accessed 5 May 2021
9. Emelin, V.A., Tkhostov, A.Sh.: Technological temptations of the information society: the limit of human external extensions. Voprosy filosofii (Prob. Philos.) (5), 84–90 (2010)
10. Flaherty, B.E.D.: Psychophysiological aspects of space flight. J. Nerv. Mental Dis. **134**(4), 378–379 (1962)
11. Fox, S.: Cyborgs, robots and society: implications for the future of society from human enhancement with in-the-body technologies. Technologies **6**, 50 (2018). https://doi.org/10.3390/technologies6020050
12. Freud, Z. The Inconvenience of Culture. Azbuka, 192 p. (2013)
13. Babatunde, G., Emmanuel, A.A., Oluwaseun, O.R., Bunmi, O.B., Precious, A.E.: Impact of climatic change on agricultural product yield using k-means and multiple linear regressions. Int. J. Educ. Manag. Eng. (IJEME) **9**(3), 16–26 (2019). https://doi.org/10.5815/ijeme.2019.03.02
14. Global Agriculture Robot Market Size 2021 research Reports collect information useful for the extensive, technical, market-oriented, commercial study Future Growth by 2026. https://www.marketwatch.com/press-release/global-agriculture-robot-market-size-2021-research-reports-collect-information-useful-for-the-extensive-technical-market-oriented-commercial-study-future-growth-by-2026-2021-07-07. Accessed 8 July 2021
15. Goff, J.L.: Civilization of the Medieval West. Progress-academy, 376 p. (1992)
16. Gurov, O.N.: «Metaparazit» kak fenomen sovremennoy massovoy kul&#39, tury. Novosibirsk. Izd-vo NGPU, pp. 41–47 (2018)
17. Haraway, D.J.: Manifestly. University of Minnesota Press, 260 p. (2016)
18. Hu, Z., Tereykovskiy, I.A., Tereykovska, L.O., Pogorelov, V.V.: Determination of structural parameters of multilayer perceptron designed to estimate parameters of technical systems. IJISA **9**(10), 57–62 (2017)
19. Huang, M.H., Rust, R.T.: Artificial intelligence in service. J. Serv. Res. **21**(2), 155–172 (2018)
20. Industry 5.0 European Commission Directorate-General for Research and Innovation Directorate F. https://msu.euramet.org/current_calls/documents/EC_Industry5.0.pdf. Accessed 5 May 2021
21. Kumar, K.A., Aju, D.: An Internet of Thing based Agribot (IOT- Agribot) for precision agriculture and farm monitoring. Int. J. Educ. Manag. Eng. (IJEME) **10**(4), 33–39 (2020). https://doi.org/10.5815/ijeme.2020.04.04
22. Bhagawati, K., Bhagawati, R., Jini, D.: Intelligence and its application in agriculture: techniques to deal with variations and uncertainties. Int. J. Intell. Syst. Appl. (IJISA) **8**(9), 56–61 (2016). https://doi.org/10.5815/ijisa.2016.09.07
23. Kudryavtsev, P.S., Konfederatov, I.Ya.: History of physics and technology. In: Guide for Students of Pedagogical Institutes. Prosveshenie, 572 pp. (1965)
24. Awadalla, M.H.A.: Spiking neural network and bull genetic algorithm for active vibration control. Int. J. Intell. Syst. Appl. (IJISA) **10**(2), 17–26 (2018). https://doi.org/10.5815/ijisa.2018.02.02
25. Mende, M., Scott, M.L., van Doorn, J., Grewal, D., Shanks, I.: Service robots rising: how humanoid robots influence service experiences and elicit compensatory consumer responses. J. Mark. Res. **56**(4), 535–556 (2019)
26. Midson, S.A.: The cyborg and the human: origins, creatureliness, and hybridity in theological anthropology. https://www.semanticscholar.org/paper/The-cyborg-and-the-human-%3A-origins%2C-creatureliness%2C-Midson/f15471a65cee4224c8f8b554a2375c4b1229f86d. Accessed 5 May 2021

27. Milov, L.V.: The Great Russian Plowman and the Peculiarities of the Russian Historical Process. ROSSPEN, 573 p. (1998)
28. Hussain, M.W., Mirza, T., Hassan, M.M.: Impact of COVID-19 pandemic on the human behavior. Int. J. Educ. Manag. Eng. (IJEME) **10**(5), 35–61 (2020). https://doi.org/10.5815/ijeme.2020.05.05
29. Monk and Countryman. https://www.pravmir.ru/monax-i-krestyanin/. Accessed 5 May 2021
30. Vani, P.D., Rao, K.R.: Implementation of smart agriculture using CloudIoT and its geotagging on Android platform. Int. J. Eng. Manuf. (IJEM) **9**(2), 43–53 (2019). https://doi.org/10.5815/ijem.2019.02.04
31. Report of the World Commission on Environment and Development: Our Common Future (1987). https://sustainabledevelopment.un.org/content/documents/5987our-common-future.pdf. Accessed 5 May 2021
32. Sinyakina, E.G.: Psychological characteristics of the Russian peasantry in the pre-revolutionary period. In: Materials of the International Conference on the History of Psychology "V Moscow Meetings", 30 June 2009— 3 July 2009, pp. 593–604. "Institute of Psychology RAS" (2010)
33. Sustainable development goals and their features. https://rosinfostat.ru/tseli-ustojchivogo-razvitiya/. Accessed 5 May 2021
34. Terebikhin, N.M.: Sacred Geography of the Russian North, 220 p. Arkhangelsk (1993)
35. The 2019 Revision of World Population Prospects. https://population.un.org/wpp/Download/Standard/Population/. Accessed 5 May 2021
36. Viola, L.: Countryman revolt in the era of Stalin: collectivization and the culture of countryman resistance. Foundation of the first president of Russia B.N. ROSSPEN, Yeltsin, 367 p. (2010)
37. Wells, J.J.: Keep calm and remain human: how we have always been cyborgs and theories on the technological present of anthropology. Rev. Anthropol. **43**(1), 5–34 (2014)

Cluster Analysis of the Loading Time-Series with the Aim of Consistent Durability Estimation

Irina V. Gadolina(✉) and Irina M. Petrova

Mechanical Engineering Research Institute of the Russian Academy of Sciences, 4, Maly Kharitonievsky pereulok, 101990 Moscow, Russian Federation

Abstract. To ensure consistent durability estimation, the engineers should create the so-called Generalized block of loading. It is based on the loading recording and the time-share of the exploitation modes during the service. In this paper, for the first time in engineering, the building of the Generalized loading block for the machine parts is considered from the mathematical point of view. Here we can see an innovative combination of techniques from cluster analysis of the time series and the specific problem of longevity estimation. This problem is important in estimating the agriculture machines longevity and reliability. To solve the problem of distinguishing the different service modes and estimating their share-time, the authors applied a popular machine learning tool, namely, cluster analysis. Because the service loading is un-stationary, selecting the modes is an important decision-making problem: the modes list should be neither too long nor too short. Because the generalized block of loading is created for the longevity estimation, the employed parameters are connected with process traits influencing fatigue damage accumulation. The semi-empirical modelled process, partly based on real loadings in service, was investigated for the case study. At the final stage, the so-called "teacher" was invoked; the result was checked using prior information.

Keywords: Random loading · Generalized block · Longevity · Agriculture machinery · Machine-learning · Cluster analysis

1 Introduction

Metal fatigue is a severe degrading process that is almost inevitable in machines part and restricts their longevity. For longevity estimation, it is important to consider the loading during service properly. The task under consideration is creating a Generalized block of loading for fatigue damage estimation [1]. There were no scientifically baked up decisions for this important engineering problem so far. The Generalized block should reflect the service modes in their proper time-share. Due to a great demand in engineering, this question has been considered in many works. For example, in [2], the authors review applying the new technical possibilities (GSP, Big Data) to loads estimation in the auto industry. The authors proposed to consider the loading variation in some particular service conditions and also make general conclusions concerning the car park loading in a whole.

Z. Hu et al. (Eds.): AIPE 2021, LNDECT 119, pp. 14–24, 2022.
https://doi.org/10.1007/978-3-030-97064-2_2

This paper proposes selecting the sub-sequences by clustering out of the extensive time recording of the process. This topic was discussed in [3]. Partly due to appearing Big data instruments, the possibility of cluster analyses for sub-sequence identification looks promising [3]. The problem of selecting the time sub-series for the load assumption in the problems fatigue longevity estimation has never been considered earlier from the mathematical perspective. Moreover, the existed methods were not computerized. In this paper, the cluster analysis was applied so that it allows us to determine the similarities and differences between the specific data from tensor gages, partitioning the sub-sets of time-series to guarantee, that the similar data are placed to a specific group or the cluster.

Cluster analysis is generally a non-supervised multivariate statistical procedure that collects information about a sample of objects and then arranges the objects into relatively homogeneous groups. It is part of data-mining domain studies [4]. In [5], problems with the popular k-means cluster method are considered, and the enhanced method of the K-means algorithm, including the computation of the weighted mean to improve the centroids initialization, is proposed. In [6], the improved K-Means algorithm was developed, and it proves that the new method of selecting initial seeds is better in terms of mathematical computation and reliability. The authors of [7] proposed the use of Simulated Annealing, a metaheuristic search algorithm, to select optimal K in K-means. Before clustering, outlier detection is required [8].

It might be stated that the basic technology of agricultural machinery has changed little over the past century. While modern harvesters and seeders may work better or differ slightly from their predecessors, today's most expensive harvester still cuts, threshes and separates grain the same way it always does. On the other hand, technology is changing the way people operate machines partly due to computer monitoring systems, GPS (accurate worldwide navigation and geodetic system based on the reception of signals from many orbiting satellites) [9].

Machine learning algorithms can be integrated into decision support tools and are increasingly popular and powerful in many areas, including precision agriculture.

Precision farming [10] began when GPS signals were made available to the manufactured and service engineers as a modern tool. Precision Farming pro-vides vehicle management as well as site-specific. Combined with telematics and data management, Precision Farming improves work accuracy and manages variations in the field. To do monitoring and control, it is necessary to collect, analyze and store a large amount of digital data (Big data). Due to the application of new information technologies, the importance of computerized tools like statistical bootstrap was stressed in [11] in increasing the productivity of wheat in Egypt. Modern deep learning algorithms, some of which are beginning to be used in precision farming [12], can make predictions based on small clusters of image functions.

To consider varied modes, there is a need for special tools. The authors of [13] created the so-called composite spectrum to reduce the number of spectra analyzed, simplifying the analysis process for machine monitoring and fault detection. In this work, vibration signals from five components of a combine harvester (thresher, chopper, straw walkers, sieve box, and engine) were recorded by placing four accelerometers along with the combine-harvester chassis. In [14], the estimation method of an expert system to predict the statuses of several agro-industrial machine rotary components by using a vibration signal acquired from a single point of the machine and a learning method to fit the estimation method. Both methods belonged to the machine-learning tools and were evaluated in an agricultural harvester.

The authors of [13] operated with the spectral density spectrum, which is believed not to be connected with longevity directly. The spectral density spectra might be helpful with diagnostic, especially with an early prognosis of failures. For longevity estimation of the machine parts, the amplitude spectra, which result from the Rainflow cycle-counting, are far more informative [15, 16]. Following the abovementioned, in this paper, the authors operate with the distribution of Rain-flow amplitudes [1, 16].

Given the difficulties associated with determining operating modes under mechanical loads and calculating the corresponding forces, the design of agricultural machines is far from a deterministic science, since variations matter [17]. This is where researchers need machine learning tools again.

The time has come to extend some of new machine-learning tools to reliability estimation of agriculture machines.

2 Peculiarities of the Agriculture Machines Loading

The knowledge of the loading modes is essential for ensuring agricultural machines reliability. The timing of equipment availability also depends on its reliability. The reliability of the agricultural machines is considered, for example, in [18], where it was stated that "one of their most important operational characteristics is reliability, i. e. the ability to perform specified functions, keeping over time the values of the operational indicators within the required limits". The availability factor with considering its variability, has been studied, in particular [19]. There are very narrow and precise time windows for every crop sector for planting and harvesting when equipment must work. Losing the equipment during those key time windows could prevent planting during the optimal planting window or delay harvest resulting in the loss of crop quality. Therefore, it is required to improve reliability and the methods of longevity estimation.

The author of [1] developed the methods for estimation longevity in a probabilistic sense. Because the loading process is non-stationary, special approaches were developed [1, 20]. Many years ago, in IMASH RAS, Moscow, researches concerning the reliability of the newly developing (in that time) developed combine were conducted. The loading of the combine harvester frame was investigated. While comparing the loading modes during the history of exploitation, the engineers revealed that due to significantly higher loading amplitudes during the machine's move by the cross-country road, most of the damage accumulation occurred exactly in this mode. The damage accumulation, estimated by the correct-ed linear hypothesis [1], showed that the engineer should consider only the amplitude distributions in the mode of moving by cross-country road to make the sound decision about the Generalized loading block. On the other hand, to judge the longevity, the researchers should have stressed out that the machine's moving regime takes only a tiny part of the general time of service (about 8%, see Fig. 1). During testing, the scientists could have been using only those data without losing prediction accuracy.

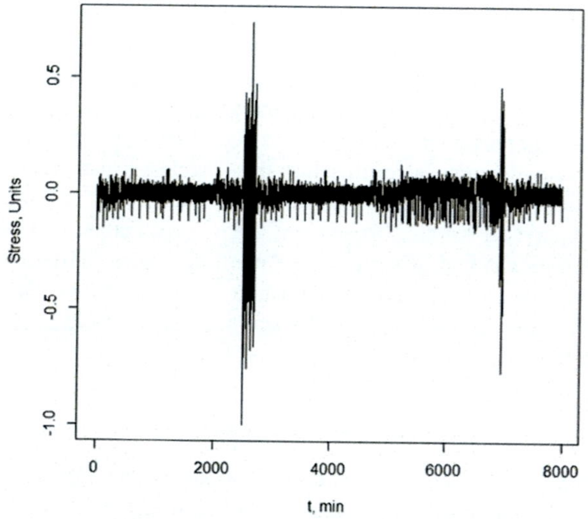

Fig. 1. Time recording of the stress during exploitation of the agriculture machinery (shown schematically).

Looking at Fig. 1, it becomes evident that only in two short pieces of realization, the stress fluctuations are significant. If one recalls the fatigue curve [1], where the exponential longevity dependence on stress is formulated, it became clear that extracting the almost zero realization parts will not lead to any noticeable change in fatigue damage. This fact again points out the importance of creating Generalized block thoughtfully.

3 Method

According to the Rainflow cycle counting procedure [1, 16] and other cycle counting methods (extremum method, range, etc.), the realization of the random loading process priorly is underdone to some preliminary treating discretization and local extremums extraction. Instead of a continuous random process, the sequences of the local maximum and minimum are investigated. It follows from the lows of fatigue damage accumulation [1]. All the forthcoming considerations are about this sequence of extrema of random processes.

The paper presents an innovative combination of techniques from 1) cluster analysis of the time series [3] and 2) the engineering problem of longevity estimation [1]. The most popular cluster-solving method, namely, K-means, has been chosen to selecting the service modes. Researchers released the K-means algorithm decades ago, and many improvements have been made since then. K -means Clustering is an unsupervised learning algorithm that tries to cluster data based on their similarity. Varied dimensions (K) tasks might be considered.

Unsupervised learning means that there is no outcome to be predicted, and the algorithm tries to find patterns in the data. In K- means clustering, we have to specify the number of clusters we want the data to be grouped into. The algorithm randomly assigns each observation to a cluster and finds the centroid of each cluster. Then, the algorithm iterates through two steps:

- Reassign data points to the cluster whose centroid is closest.
- Calculate the new centroid of each cluster.

These two steps are repeated till the within-cluster variation cannot be reduced any further. The within-cluster variation is calculated as the sum of the Euclidean distance between the data points and their respective cluster centroids.

This method is similar to the reverse analysis of variance (ANOVA) method because the test of significance in the analysis of variance compares the between-group variability with the within-group variability when testing the hypothesis that the means in the groups differ from each other. Once the K-means cluster analysis results are obtained, the means for each cluster can be calculated for each dimension to estimate how the clusters differ from each other.

The procedure of extracting of sub-sequences out of the initial (long) sequence of extremums is described as follows:

(1) The process under investigation is evenly divided by equal parts by length (those parts are called "sub-sequences" later on);
(2) The most important for fatigue estimation (and fatigue damage accumulation) parameters are estimated for each sub-sequence (if some of the estimated parameters are strongly correlated, it might be considered to exclude some of them as not being highly informative to reduce the number of parameters;
(3) The resulting matrix (z x n) is underdone by the K-means cluster method. z is the number of parameters, and n is the number of sub-sequences.
(4) Although the cluster analysis belongs to the group of the machine-learning tool "without the teacher," the teacher was revealed at the final stage to prove the workability of the proposed method.

4 Case Study. The Process Under Investigation

A semi-empirical random R1 process was created based on well-known SAE loading processes [21]. With the aim of the investigation, certain parts of the three processes were randomly mixed and later divided into equal short sub-sequences. As stressed

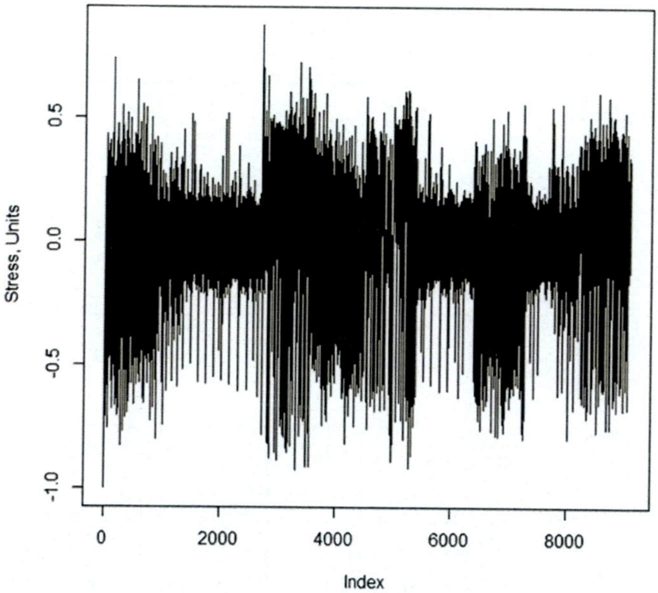

Fig. 2. Semi-empirical random R1 process built on the base of SAE [21]. loading processes

before, the process consists only of extremum values: MIN-MAX-MIN-MAX … arranged in the proper sequence. The process R1 in condensed form is shown in Fig. 2:

For the study, we composed the process R1 out of varied SAE processes: Bracket, Suspension, Transmission [21]. The goal of the investigation was to prove the possibility of identification of those varied regimes in an overall random realization. Bracket, Suspension, Trans-mission processes were transformed, in such a manner, that their asymmetry became close to minus one (−1) (see Fig. 2). The clustering was performed in multi-dimensional space. The initially selected parameters related to longevity with the selected estimated values are shown in Table 1.

Table 1. Head (upper rows) of the Table of factors with their sample values

	RMS	Samax	V	I	R	f
Sub-realization number	Root mean square [Units]	Maximum amplitude in sub-realization [Units]	Fullness factor [−]	Irregularity factor [−]	Asymmetry Smin/Smax [−]	Frequency [Hz]
1	0.32	0.870	0.560	0.90	−1.049	10
2	0.30	0.610	0.660	0.96	−0.882	10
3	0.29	0.640	0.790	0.89	−0.927	10
4	0.21	0.597	0.460	0.89	−1.385	12
5	0.18	0.485	0.472	0.87	−0.777	12
…	…	…	…	…	…	…
30	0.28	0.67	0.53	0.73	−0.65	15

As it was mentioned in paragraph 3.2. at the first step, the pair correlations among the factors were investigated. Their graphical representation of the factors is shown in Fig. 3 (Graphical representation here and further was performed in the R-programming package [22]).

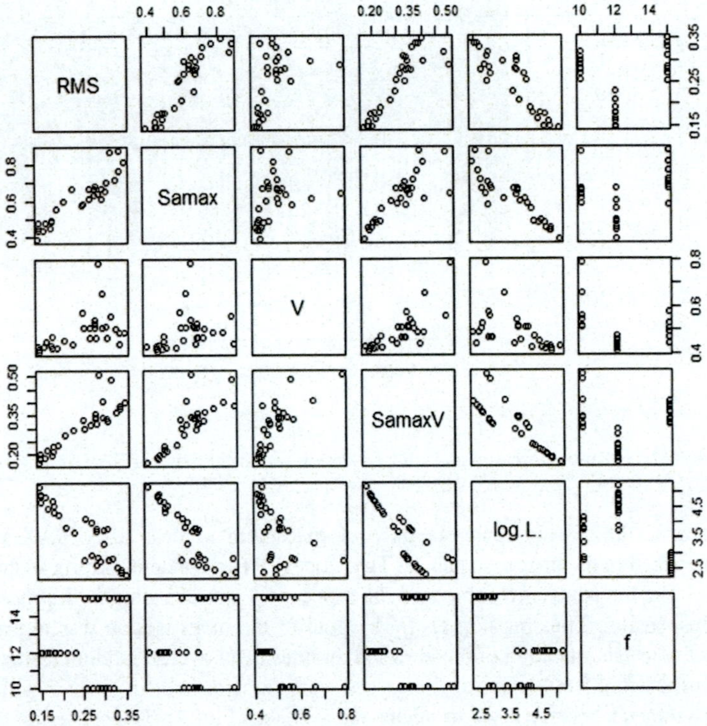

Fig. 3. Pair correlations of the selected parameters of the sub-realization of the process R1

As can be seen from Fig. 3, the parameters RMS and Samax are strongly correlated (cor [RMS, Samax] = 0.92), which allows us to exclude one of them. We decided to leave Samax as it being proved more suitable for fatigue calculations. Equivalent amplitude Se = V* Samax correlate (negatively) strongly with the estimated longevity L(cor[Se,L] = − 0.89). Because L was the estimated value, we decided on leaving Se as more related to the raw recording.

Also, as preliminary analysis before clustering, the 3d representation of three influential parameters is shown in Fig. 4 (3d has been chosen for more understandable visualization):

It might be seen from the picture that there are three distinct groups in the data, which would be proved later on.

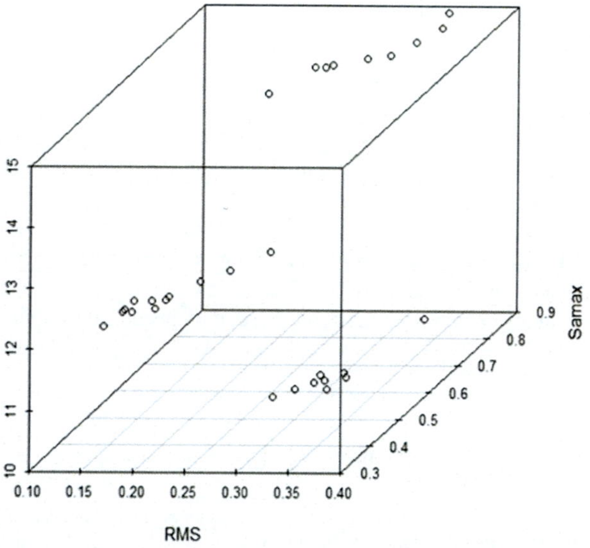

Fig. 4. 3d image of the parameters space of the sub-realizations of R1

5 Results and Discussion

After 10 trials of clustering iterations, K-means clustering program in R language [22] estimated 3 clusters of sizes 9, 9, 12 with cluster means as follows (Table 2):

Table 2. Cluster analysis centers in R1 process investigation

Cluster index	RMS	Samax	V	I	R	f
1	0.306	0.720	0.493	0.66	−1.018	15
2	0.280	0.658	0.568	0.91	−1.049	10
3	0.174	0.494	0.435	0.92	−0.985	12

The K-mean clustering program from R-package [22] estimated statistics as follows:

Within cluster sum of squares by cluster:

$$0.1930065\ldots\ldots0.2704412\ldots..0.9040177$$
$$(between\_SS/total\_SS = 98.8\%)$$

The F-statistic values obtained for each dimension are another indicator of how well the corresponding dimension discriminates against clusters. However, the F -tests should be used only for descriptive purposes here because the clusters have been chosen to maximize the difference among clusters. The observed significance levels are not intended for hypothesis proving.

For confirmation of the workability of the developed method, at the next step, we employ the "checking with a teacher", although it does not well correspond to the idea of clustering. As a matter of fact, we possessed the prior information concerning the sets belonging to particular groups. The matching of data is satisfactory (Table 3). The cluster analysis selected 3 clusters, and they corresponded well to the initial data. All the parts from "Bracket" [21] were identified as cluster #2, "Suspension" as cluster #3 and "Transmission" as cluster #1. Such a good match verifies that the choice of parameters for distinguishing the loading modes was correct and the method based on cluster analysis proceeds the random process peculiarities well.

Table 3. Cluster statistic and its belongings

Sets' belongings to the cluster, selected by program:																													
2	2	2	3	3	3	3	3	3	1	1	1	2	2	2	1	1	1	3	3	3	2	2	2	3	3	3	1	1	1
Prior information of sets' belongings (hidden "teacher"):																													
Bracket	Bracket	Bracket	Suspension	Suspension	Suspension	Suspension	Suspension	Suspension	Transmission	Transmission	Transmission	Bracket	Bracket	Bracket	Transmission	Transmission	Transmission	Suspension	Suspension	Suspension	Bracket	Bracket	Bracket	Suspension	Suspension	Suspension	Transmission	Transmission	Transmission

The proposed method successfully distinguishes the mixed modes "Bracket", "Transmission" and "Suspension" – see Table 3. From this we can conclude, that the method will work in the situations with real stress recordings.

In connection with the problem under discussion [23], one should point out to the fundamental difference between cluster analysis and grouping tasks. If the classes are real, natural, exist in reality, clearly separated from each other, then any cluster analysis algorithm will distinguish them. Consequently, as a criterion for the naturalness of the classification, stability concerning the choice of the cluster analysis algorithm should be considered. Stability can be tested by applying several approaches to the data, for example, such dissimilar algorithms as "near neighbors" and "far neighbors". If the results obtained are substantively close, then they are adequate to reality. Application of the other dissimilar algorithms by the authors of this paper is planned in future studies.

6 Conclusions

The actual problem of selecting the varied loading modes for the creation a Generalized block of loading for fatigue studies has been solved with cluster analysis, which belongs to a group of machine learning tools. This is the first attempt to solve this problem systematically. Due to the widespread expansion of new devices in agriculture, developing the measuring technic and newly created methods for information processing, it is important to develop new methods for longevity estimation, collecting information about the loading modes.

The developed method allows selecting a reasonable number of the loading modes to generate information about loading during the life cycle. It is expected to bring more valuable information about the loading of the machines' part during service. That information will allow improving longevity estimation precision. The result of the performed investigation has been proved by comparing it with the prior information (which was hidden at the beginning with the investigation aims). The method is expected to work well in cases of loading investigation in the parts of the movable agricultural machines. Based on the data, which are expected to appear after the publication of this paper from its readers, the authors hope to perform the future examination of the method and application to the fatigue damage investigation of the machine's part.

Acknowledgements. Most of the calculations, as well as the modelling, were performed in freely-distributed R-language package [22].

References

1. Kogaev, V.: Calculations for strength for variables in time strains (1993). (in Russian)
2. Burger, M., Dressler, K., Speckert, M.: Load assumption process for durability design using new data sources and data analytics. Int. J. Fatigue **145**, 106116 (2021)
3. Aghabozorgi, S., Shirkhorshidi, A.S., Wah, T.Y.: Time-series clustering – a decade review. Inf. Syst. **53**, 16–38 (2015)
4. Al-Hagery, M.A., Alzaid, M.A., Alharbi, T.S., Alhanaya, M.A.: Data mining methods for detecting the most significant factors affecting students' performance. Int. J. Inf. Technol. Comput. Sci. (IJITCS) **12**(5), 1–13 (2020). https://doi.org/10.5815/ijitcs.2020.05.01
5. Fahim, A.: Finding the number of clusters in data and better initial centers for K-means algorithm. Int. J. Intell. Syst. Appl. (IJISA) **12**(6), 1–20 (2020). https://doi.org/10.5815/ijisa.2020.06.01
6. Fabregas, A.C., Gerardo, B.D., Tanguilig, B.T., III.: Enhanced initial centroids for K-means algorithm. Int. J. Inf. Technol. Comput. Sci. (IJITCS) **9**(1), 26–33 (2017). https://doi.org/10.5815/ijitcs.2017.01.04
7. Onyezewe, A., Kana, A.F., Abdullahi, F.B., Abdulsalami, A.O.: An enhanced adaptive K-nearest neighbor classifier using simulated annealing. Int. J. Int. Syst. Appl. (IJISA) **13**(1), 34–44 (2021). https://doi.org/10.5815/ijisa.2021.01.03
8. Sakr, M., Atwa, W., Keshk, A.: Genetic-based summarization for local outlier detection in data stream. Int. J. Intell. Syst. Appl. (IJISA) **13**(1), 58–68 (2021). https://doi.org/10.5815/ijisa.2021.01.05

9. Public-Private Analytic Exchange Program. Threats to Precision Agriculture (2018). https://www.researchgate.net/publication/339052593_Threats_to...
10. Kamilaris, A., Boldú, P.: Deep learning in agriculture: a survey. Comput. Electron. Agric. **147**(1) (2018). https://doi.org/10.1016/j.compag.2018.02.016
11. Khalaf, A.E.A., et al.: Development of a five-parameter model to facilitate the estimation of additive, dominance, and Epistatic effects with a mediating using bootstrapping in advanced generations of wheat. Agronomy, **11**, 1325 (2021).https://doi.org/10.3390/agronomy11071325
12. Celenta, G., De Simone, M.C.: Retrofitting techniques for agricultural machines. In: Karabegović, I. (ed.) NT 2020. LNNS, vol. 128, pp. 388–396. Springer, Cham (2020). https://doi.org/10.1007/978-3-030-46817-0_44
13. Feijoo, F., et al.: Application of composite spectrum in agricultural machines. Sensors, September 2020.https://doi.org/10.3390/s20195519
14. Martínez, V., Gomez-Gil, J., Ruiz-Gonzalez, R.: An artificial neural network based expert system fitted with genetic algorithms for detecting the status of several rotary components in agro-industrial machines using a single vibration signal/Expert Syst. Appl. **42**, 6433–6441 (2015)
15. Gadolina, I.V., Makhutov, N.A., Erpalov, A.V.: Varied approaches to loading assessment in fatigue studies. Int. J. Fatigue. **144**, 106035 (2021). https://doi.org/10.1016/j.ijfatigue.2020.106035
16. Endo, T., et al.: Damage evaluation of metals for random or varying loading—three aspects of rain flow method. Mech. Behav. Mater. **1**, 371–380 (1974)
17. Abo Al-kheer, A., et al.: Integrating optimization and reliability tools into the design of agricultural machines. In: Proceedins of the 20ème Congrès Français de Mécanique, Besançon, 29 août au 2 septembre 2011
18. N.S., Shukhanov, Kuzmin, A.V.: Reliability of machine-tractor aggregates operation. Eng. Technol. Syst. **30**(1):8–20 (2020). https://doi.org/10.15507/2658-4123.030.202001.008-020 (in Russian)
19. Papich, L., et al.: Interval estimation of the availability factor of the bucket-wheel excavator based on bootstrap modeling. J. Mach. Manuf. Reliab. **45**(6), 531–537 (2016). https://doi.org/10.3103/S1052618816060091
20. Gadolina, I., Zaynetdinov, R.: The estimation of the sufficient random loading realization length in the problem of machine parts longevity. In: Proceedings of 2018 IEEE 9th International Conference on Dependable Systems, Services and Technologies, DESSERT 9, pp. 159–162 (2018)
21. Tucker, L., Bussa, S.: The SAE Cumulative Fatigue Damage Test Program. In book Fatigue under Complex Loading. Ed. R.M. Wetzel, pp.1–44. Society of Automotive Engineering (1977)
22. Team, R.C.: A language and environment for statistical computing. Found. Stat. Comput. (2020)
23. Orlov, A.I.: Econometrics. Exam Publishing House Textbook, Moscow (2002)

Intellectual Biotechnical Systems in Livestock: Theoretical Aspects

Vladimir Kirsanov Vyacheslavovich, Yurij Tsoj Alekseevich, and Daria Geletiy Grigorievna(✉)

Federal State Budgetary Scientific Institution,
«Federal Scientific Agroengineering Center VIM», Moscow, Russia
geletiyd@gmail.com

Abstract. Biotechnical systems belong to the class of man-machine systems or "man-machine-animal" systems, the latter belong to the livestock industry. Biotechnical systems in animal husbandry have the properties of bimodality, when there are two or more biological objects, a person as a managing operator and a service object (plants, animals). To increase the efficiency of the functioning of complex biomachine and technical systems, it is necessary to conduct systemic studies for further intellectualization and digital transformation. There are two approaches in the study of human-machine systems: anthropocentric and machine-centric. Functionals of the subsystems "Human" and "Machine" are considered, while part of the functions of the human operator (HO) will be gradually given to the "Machine" (M), and (HO) will be transformed into a human-expert (HE) and a human-user (HU). A scheme for the functioning of local BTS in a partially autonomous mode of multi-agent control has been developed, performance criteria for local biotechnical systems LBTS have been determined.

Keywords: Biotechnical system · Biomachsystem · Local biotechnical system · Levels of adaptation · Machine-centric model · Human operator (HO) - user (HU) - expert (HE)

1 Introduction

There is a class of man-machine systems, which is transformed into the system "man-machine-plant", "man-machine-animal" in agriculture [1]. As you can see, in agricultural production, biotechnical systems have the properties of bimodality, when there are two or more biological objects, a person as a managing operator and a service object (plants, animals). In addition, the uniting environment of these biological objects is also a biological object: the field (soil) - as the main source of energy for plants and animals. The process of energy conversion is considered in agricultural production in the classical works of Academician V.P. Goryachkin, where he considers three components of this process: an energy source (sun, atmosphere), energy storage (soil as accumulator), energy receiver (consumer) - plants, animals [2]. These works were invaluable for the progress of the theory and practice of creating new technology for agricultural production, converting its designing to scientific basis. The increasing complexity of machine technologies led to the formation and development of scientific

Z. Hu et al. (Eds.): AIPE 2021, LNDECT 119, pp. 25–33, 2022.
https://doi.org/10.1007/978-3-030-97064-2_3

methods of substantiating and calculating them, not only individual working elements were analyzed, but also aggregates at large, which affected at the quality of service and the state of biological objects, ecology, also a cluster of a man-machine systems was formed [2]. In animal husbandry for biotechnical systems is used the term BTS. This term is used in Professor L.P. Kartashov works, when he was studying the process of interaction of milking equipment mechanisms with a cow's udder [3]. There are known works on functional systems in the field of medicine, performed by P.K. Anokhin [4]. Professor V.M. Akhutin (1975) suggested the term "Biotechnical systems" (BTS) to distinguish a special class of large systems, which are a combination of biological and technical elements interconnected in a single control loop [5]. Currently, under the leadership of Academician of the Russian Academy of Sciences V.I.Chernoivanov, research is being carried out in this direction, a new concept "Biomashsystem" has been introduced, which in agricultural production more accurately reflects the interaction of machines and units with biological objects [6].

The term "Equipment" is often used in animal husbandry. The traditional self-moving machines include feed preparation and dispensing units, therefore, both terms, BTS and "Biomashsystems", are obviously applicable here. With the development of automation and informatization of production, the role and importance of systematization of knowledge and management of complex biotechnical and machine complexes increases, especially in such an important industry as animal husbandry.

2 Purpose of Research

To improve the efficiency of complex biomachine and technical systems, it is necessary to develop their system research in order to further intellectualize and digitalize agricultural production [7].

3 Theoretical Aspects of a Research

In modern milking parlors operator's work is quite significant, he has to make frequent movements to service animals, rub off the dug, milk the first streams, connect milking machines for a long time 8–10 h per shift. All these operations cause physical fatigue and information overload. Such identical, monotonous operations will undoubtedly be performed by robots in the future. The only question is the high cost, today, of robotic technologies. In this regard, an analysis of the ergonomics of the work of milking machine operators involved in high-performance milking stations deserves attention. The higher the labor productivity, the higher the flow rhythm and the frequency of movements performed by the operator.

There are two approaches to the study of human-machine systems: anthropocentric and machine-centric [8]. The first assumes and assigns a decisive role to the person, the second to the machine. Of course, the role of "Man" in the system "man-machine-animal" is not only preserved, but also increases. However, with the development of automation tools, the functions of the human operator "HO" will gradually be replaced by a human-expert "HE" and a human-user "HU".

The general functionality of the "human" subsystem can be represented as the sum of the functionals:

$$Z_h = f[\mathrm{HO}] + f[\mathrm{HE}] + f[\mathrm{HU}] \quad (1)$$

herewith, f [HO] tends to a minimum in automatic systems, while the role of accumulated human knowledge in models and algorithms increases and they transfer to the functionals f [HE] and f [HU] [9].

Video timing table for milking rotaries (Table 1).

Table 1. Timing of manual operation on rotary milking parlor

№	Rotary milking parlor	Number of operator movements per 1 cow	Duration of manual operations, s	Energy intensity of operations per 1 cow, J	Note
1	Milkline (operator outside)	8–9	5–6 - cleaning dug before milking	3,92	1st operator
		6	8–9 - milking machine connection	20,59	2 operator
		5	4–5 - dug treatment after milking	3,92	3rd operator
2	AutoRotor 40 (operator inside)	14–15	12–13 - cleaning dug before milking and milking machine connection	37,26	1st operator
		14–15	12–13c - cleaning dug before milking and milking machine connection	37,26	2nd operator
		4–5	4–5 - dug treatment after milking	3,92	3rd operator
3	AutoRotor 40 (outside)	8–9	5–6 cleaning dug before milking	3,92	1st operator
		6	6–8 - milking machine connection	20,59	2nd operator
		4–5	4–5 - dug treatment after milking	3,92	3rd operator
4	AutoRotor 60 (operator outside)	12–13	10–12 - cleaning dug before milking and milking machine connection	24,51	1st operator
		12–13	10–12 - cleaning dug before milking and milking machine connection	24,51	2nd operator

4 Possible Application

Therefore, at the first stage, the manual control functions will be transferred to the machine, and the intelligent functions of the operator will gradually be transferred to the machine (automaton), while the person will remain in control and correction of machine algorithms, animal behavior models, etc. The functionality of the "machine" will increase.

$$Z(\mathrm{m}) = f[\mathrm{HO}, x_{1\mathrm{m}} \ldots x_{n\mathrm{m}}] + f[\mathrm{HO}, y_{1\mathrm{i}} \ldots y_{n\mathrm{i}}] + f[\mathrm{M}, z_{m1} \ldots z_{mn}] + f[\mathrm{A}, y_{\mathrm{A}1} \ldots y_{\mathrm{A}n}] \quad (2)$$

$Z(\mathrm{m})$ – full functionality of the machine;

$f[\mathrm{HO}, x_{1\mathrm{m}} \ldots x_{n\mathrm{m}}]$ – the functionality of the replaced manual operations of the human operator transferred to the machine (automaton);

$f[\mathrm{HO}, y_{1\mathrm{i}} \ldots y_{n\mathrm{i}}]$ – the functional part of the simple intelligent functions (HO) transferred to the machine (M) (control, analysis, operation) with the exception of intuition;

$f[\mathrm{A}, y_{\mathrm{A}1} \ldots y_{\mathrm{A}n}]$ – functional parameters of animals previously controlled by a human operator: control of milk flow, heat, mastitis incidence, etc., transmitted to the machine.

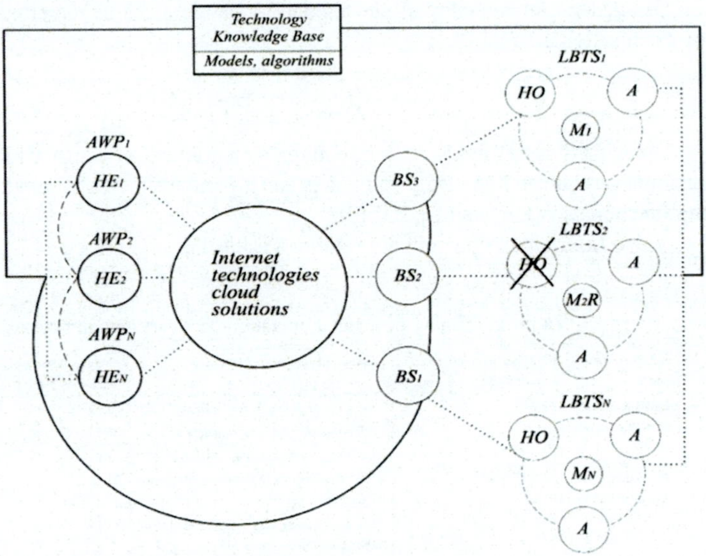

Fig. 1. Block diagram of an intelligent BTS

In this context, the role of the "machine" factor in modern agricultural technologies will increase [10]. For a person will be mainly assigned the functions of "HU" and "HE". Based on the "weak" neural network artificial intelligence, the machine will gradually learn relatively simple functions of adaptation to biological objects: (the quantity and quality of feed consumed, the measurement of milk yield and the completeness of milking, disease control and treatment). At the first stage of production intellectualization, the information component will increase, illustrating the interaction of the "machine-animal" subsystem. And the machine will use its sensors to observe and signal problems in the subsystem (M-A), what went wrong: decreased milk yield, acidosis of the rumen of the cow's stomach, etc. The signals will be transmitted through the corresponding base stations to the automated workstations of specialists (Fig. 1) to the "human expert": a specialized specialist (veterinarian, zooengineer, etc.), who will study them, compare them, and transmit commands via feedback to local biotechnical and biomachine systems of the LBTS (BMS), redirect and increase the levels of their adaptation (*Ca*).

The levels of adaptation of local biotechnical systems (milking, drinking, feeding, etc.) can be estimated by appropriate coefficients.

$$\begin{cases} Y_{a1} = \dfrac{N_{a_1}}{N_{\sum_1}} \\ Y_{an} = \dfrac{N_{a_n}}{N_{\sum_n}} \end{cases} \tag{3}$$

N_{a_1}, N_{a_n}- the number of indicators in the subsystem (M-A) that are controlled automatically and do not require human intervention,

$N_{\sum 1}, N_{\sum n}$ - the total number of indicators that require monitoring for the normal operation of the LBTS, including by specialists (HE, HU).

At the same time, the operation of individual LBTS may not require communication with the "control center" (HE, HU) through the corresponding AWP to make a decision, and directly transmit signals from one LBTS to another, using autonomous multi-agent control modes [11]. For example, the $LBTS_f$ of feeding receive a signal from the $LBTS_m$ of milking about the changed conditions of functioning - decreased milk yields in a group of lactating animals, without changing their health indicators. The LBTS of feeding system makes a decision to adjust the feeding rations by sending a duplicate signal to the zootechnician.

In this case, only the I local control loop can work without "entering" the (DB) and (AWP). It is enough to adjust the ration automatically, the knowledge of specialists is not required, they can only send SMS messages in the form of $(I^M_{a_i}, I^F_{a_i})$. Other LBTS (microclimate, manure removal, etc.) can work similarly (Fig. 2).

For example, the parameters of the microclimate change, the gas content of the room increases (NH3, CO2, H2S) when the manure removal system is working, the forced ventilation system turns on, which removes excess harmful gases.

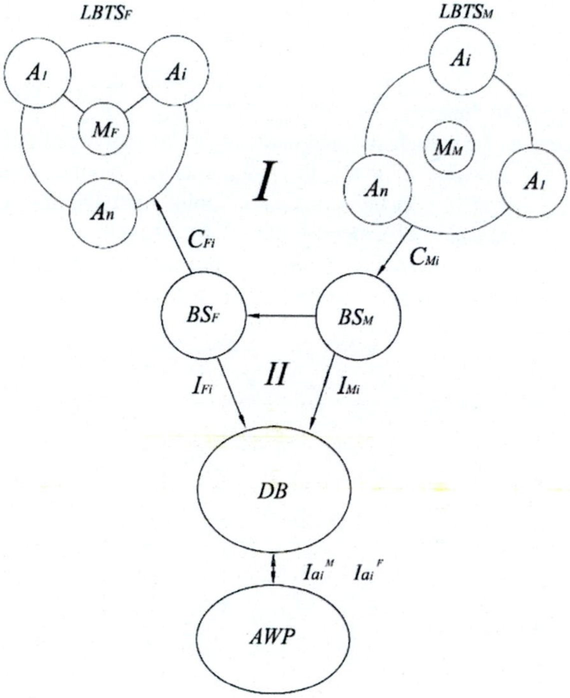

Fig. 2. Operation of local BTS in partially autonomous multi-agent control mode

BS_F, BS_M – base stations that transmit the signal to the LBTS feeding and LBTS milking, respectively.

$C^M_{a_i}$, $C^F_{a_i}$, – accordingly, the control adaptation signals from (LBTS) m, (LBTS)f.

I^M_a, $I^F_{a_i}$, – information signals transmitted through the DB (databases) to the AWP (HE, HU).

In case when intervention of specialists is required and the local level of interaction cannot correct the situation, information signals pass to the automated workplaces AWP (HE, HU) and specialized specialists make a decision, examine animals in separated groups allocated by the subsystem. A block of test parameters is formed in each local subsystem. Deviations of the current parameters of biological and machine objects are measured and compared with their test characteristics and limit values [12].

$$\begin{cases} \Delta[\mathrm{OI}_{A_1\ldots A_N}]^C \le \Delta[\mathrm{OI}_{A_1\ldots A_N}]^{per} \\ \Delta[X_{1_p\ldots n_p}]^C \le \Delta[X_{1_p\ldots n_p}]^{per} \\ \Delta[z_{M_1\ldots M_N}]^C \le \Delta[z_{M_1\ldots M_N}]^{per} \end{cases} \tag{4}$$

$\Delta[\mathrm{OI}_{A_1\ldots A_N}]^C, \Delta[\mathrm{OI}_{A_1\ldots A_N}]^{per}$ – respectively, the current and permissible deviations of the controlled indicators of biological objects (animals);

$\Delta[X_{1_p\ldots n_p}]^C, \Delta[X_{1_p\ldots n_p}]^{per}$ – respectively, current and permissible deviations in the operating modes of the human operator (HO);

$\Delta[z_{M_1\ldots M_N}]^C, \Delta[z_{M_1\ldots M_N}]^{per}$ – respectively, current and permissible deviations in the operation of the “machine factor”.

Because of the complexity and diversity of the monitored indicators, this grouping will make it possible to study them separately. The indicators of the first group $\Delta[\mathrm{OI}_{A_1\ldots A_N}]$ can include lactation indicators (milk yield, Qi, milk yield rate, Vi, milking duration t_{d_i}, indicators of general growth (animal body weight, the ratio of fat, bone, muscle mass, animal size, its appraisal characteristics, etc.) [13].

The indicators of animal health can include the number of somatic cells in milk, abnormalities in growth of reproductive organs, limbs, etc.

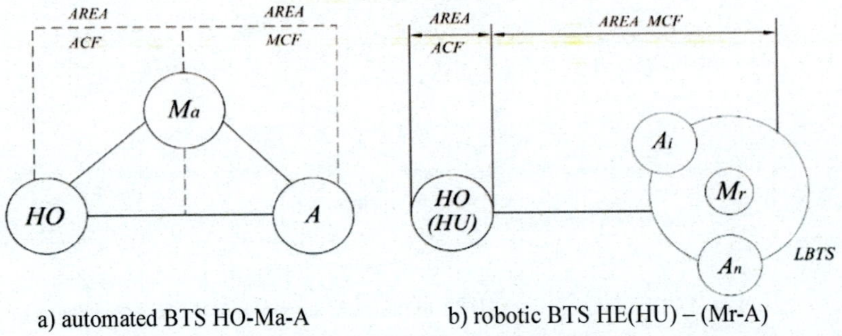

Fig. 3. The process of converting an automated BTS into a robotic one

The second group of indicators characterizing the work of a human operator (HO) may include the following indicators: duration of manual operations for servicing an animal, t_{mo}, intensity of manual operations performed per unit of time n_m; energy consumption of the performed operations, E_m.

The group of indicators characterizing the "machine" factor includes the duration of the servicing cycle of animals, the consumption characteristics of the distributed, received materials (feed, manure, milk, etc.). Thus, at the present stage of the development of BTS in animal husbandry, a three-link system is being transformed into a two-link system with the polarization of subsystems (HE, HU) and subsystems (M-A) (Fig. 3) [14, 15].

There is a polarization of the subsystem (M-A) into a local machine-centered model of the LBTS, which can function with a high degree of autonomy, while the "powers" of the "HO" are delegated to the machine, the zone of "influence" of the anthropocentric factor (ACF) decreases, but at the same time the level of its intellectualization increases, since a number of manual operations performed earlier by the "HO" are handed to the machine. The level of intellectualization L_i should not be mistaken for the level of automation L_a. Since the latter replaces only simple mechanical operations with manual control. And the level of intellectualization includes the level of automation plus additional visual and analytical functions of the human operator, which he always used for evaluating the interaction of the subsystem (M-A).

Thus, the formalization of the level of intellectualization of the H-M-A system can be represented by:

$$L^{I}_{H-M-A} = L^{a}_{H-M-A} \wedge f[HO, y_{1I} \ldots y_{nI}] \quad (5)$$

$L^{I}_{H-M-A}, L^{a}_{H-M-A}$ – respectively, levels of intellectualization and automation of the system (H-M-A);

$\wedge$- disjunction operator (logical addition);

$f[HO, y_{1I} \ldots y_{nI}]$ - the functional of a part of simple intelligent functions (HO) handed to machines (M).

The level of automation of the La can be calculated from the known dependence [16, 17]

$$L^{a}_{H-M-A} = \frac{Q^{a}_{H-M-A}}{Q^{a}_{H-M-A} + Q^{M}_{H-M-A}} \quad (6)$$

$Q^{a}_{H-M-A}, Q^{M}_{H-M-A}$ – respectively, quantity of automated and mechanized operations in the H-M-A system.

In an automated system (HO-Ma-A), the human operator works by directly interacting with the machine and the animal, performing known functions and manual operations. In a robotic system, some of the manual and simple intellectual functions of the HO are performed by the machine, and (HE, HU) are having functions of increased analytical content (high intelligence, prediction, etc.) [15].

The number of intelligent visual and analytical operations transmitted to the machine by a human operator is constantly growing. Therefore, the functionality that generalizes these functions is also growing rapidly

$$f = [\mathrm{HO}, y_{1_I} \ldots y_{n_I}] \rightarrow max \tag{7}$$

A human operator (HO) cannot grasp the immensity, keep track of everything, especially on a large farm, so his task as (HE, HU) is to analyze reports sent from the LBTS or individual “alarm” signals from animals, and to intervene in the process in “threatening” emergency situations related to emergencies of the machine or animal diseases, etc.

It is necessary to study separately whole groups of indicators related to the components of a complex biomachsystems, which is a modern dairy farm.

5 Conclusion

Thus, a modern automated or partially robotic dairy farm is a complex multi-level biomachsystem, including partially or completely autonomous local biotechnical subsystems that perform specific technological processes, subsystems for receiving and transmitting signals of animals, machines (base stations), an information and analytical center, which includes automated workplaces (AWP) of chief specialists who make management decisions on certain situations. The progress of biomachsystems in animal husbandry is based on the expanding machine-centric model of the subsystem (M-A), which takes in more and more intellectual functions passed on by “Human”, the latter reserves control, coordination and management of the entire system. There is a lot of computer controlled systems in animal husbandry nowadays (feeding system, water treatment, some steps of milking system such as identification or milk quality analysis). Further work is targeted on minimizing “human factor” for milking operations such as dug treatment or connecting milking machines. One of the options is to use laser scanning for dug with robotic manipulator to replace manual operations.

References

1. Chernoivanov, V.I.: Biomashsystems. Theory and applications. In: Chernoivanov, V.I. (ed.) FSBSI “Rosinformagrotech” Academician of the Russian Academy of Sciences. CIT, Moscow (2016)
2. Goryachkin, V.P.: Agricultural mechanics. In: Fundamentals of the Theory of Agricultural Machines and Tools: 1917–1918. Publishing House of Students of Petrovskaya agricultural akademie, Moscow (1919)
3. Kartashov, L.P., Solov'ev, S.A., Asmankin, E.N., Makarovskaya Z.V.: Calculation of the executive mechanisms of the biotechnical system. RAS. URB. Institute of Applied Mechanics (2002)
4. Anokhin, P.K.: Nodal questions of the theory of functional systems (1980)
5. Pepechitelev, E.P.: Problems of the synthesis of biotechnical systems. Sci. Rev. Tech. Sci. **2**, 54–62 (2016)

6. Chernoivanov, V.I., Sudakov, S.K., Tolokonnikov, G.K.: Biomashsystems, functional systems and category theory of systems. Bull. All-Russ. Res. Inst. Anim. Husbandry Mech. 2(26) (2017)
7. Arkhipov, A.G., Kosogor, S.N., et al.: Digital Transformation of Agriculture in Russia: Official edn. FBSI "Rosinformagrotech", Moscow, 80 pp. (2019)
8. Dozortsev, V.M.: Notes on order of the day and man in industrial automation. Autom. Ind. 2 (2011)
9. Doronin, A.M., Romanov, D.A., Romanov, M.A.: Human-machine interaction and its indicators. Adygeyskiy SU Bull. **4**, 244–250 (2005)
10. Lachuga, Yu.F., et al.: Strategy of machine technological modernization of agriculture for the period up to 2020, 80 pp. FBSI "Rosinformagrotech", Moscow (2009)
11. Petrin, K.V., Teryaev, R.D., Filimonov, A.B., Filimonov, N.B.: Technologies in ergatic control systems. In: Proceedings of the Southern Federal University. Technical Sciences, no. 3 (2010)
12. Kirsanov, V.V.: Methods and technical means of milk accounting and control of operation parameters at milking equipment. Ph.D. dissertation abstract. All-Russian Research Institute of Agricultural Electrification, 19 pp. (1992)
13. Karande, A.M., Kalbande, D.R.: Weight assignment algorithms for designing fully connected neural network. Int. J. Intell. Syst. Appl. (IJISA) **10**(6), 68–76 (2018). https://doi.org/10.5815/ijisa.2018.06.08
14. Dharmajee Rao, D.T.V., Ramana, K.V.: Winograd's inequality: effectiveness for efficient training of deep neural networks. Int. J. Intell. Syst. Appl. (IJISA) **10**(6), 49–58 (2018). https://doi.org/10.5815/ijisa.2018.06.06
15. Hu, Z., Tereykovskiy, I.A., Tereykovska, L.O., Pogorelov, V.V.: Determination of structural parameters of multilayer perceptron designed to estimate parameters of technical systems. Int. J. Intell. Syst. Appl. (IJISA) **9**(10), 57–62 (2017). https://doi.org/10.5815/ijisa.2017.10.07
16. Awadalla, M.H.A.: Spiking neural network and bull genetic algorithm for active vibration control. Int. J. Intell. Syst. Appl. (IJISA) **10**(2), 17–26 (2018). https://doi.org/10.5815/ijisa.2018.02.02
17. Abuljadayel, A., Wedyan, F.: An approach for the generation of higher order mutants using genetic algorithms. Int. J. Intell. Syst. Appl. (IJISA) **10**(1), 34–45 (2018). https://doi.org/10.5815/ijisa.2018.01.05

Intelligent Systems of Telemedicine Monitoring for Countryside and Agriculture

Lev I. Evelson[1(✉)], Boris V. Zingerman[2], Olga S. Kremenetskaya[3], and Nikita E. Shklovskiy-Kordi[4]

[1] Innovation Scientific Center of Information and Remote Technologies, 241050 Bryansk, Russian Federation
[2] TelePat, 196787 Moscow, Russian Federation
[3] Center for Theoretical Problems of Physicochemical Pharmacology, 119991 Moscow, Russian Federation
[4] National Medical Research Center for Hematology, 125167 Moscow, Russian Federation

Abstract. The software platform for telemedicine monitoring with Artificial Intelligence (AI) elements is presented. The principles and methods of development of the platform are described and discussed. The systems created on a base of the platform would be useful for countryside inhabitants' life quality.

Now the platform includes specific Medical Messenger, Subsystem of Intelligent Agents (IA), Subsystem of storage of the personal medical data. IA unload a doctor from routine work. The IA inform the doctor when his direct urgent attention and actions are needed. They also provide a patient and a physician with general information relevant to the patient statement picking out and consolidating it from the Patient and the Doctor Digital Libraries. Good results of the test exploitation of the systems created on a base of the platform are presented. It is displayed that objective medical indicators as well as subjective perception by the doctors and patients of RHM were positive. Already created RHM systems can be used now (and were used in the pilot running) for countryside inhabitants relating to corresponding patient category and diseases class. It is also expedient to develop specific RHM system for health safety of people living in countryside. So, there are perspective opportunities to improve social conditions of countryside and promote agriculture.The similar approach can be used also directly for agricultural stock-raising. The "televeterinary" platform for home animals had been already developed. Systems implementing remote monitoring with sensors placed on the animals and with participation of veterinary surgeons should be created. This article invites to multidisciplinary co-operation to develop and implement RHM systems for countryside and agriculture.

Keywords: Intelligent agents · Telemedicine · Remote health monitoring · Countryside · Agriculture · Artificial intelligence · Veterinary medicine

1 Introduction

Development of agricultural complex is impossible without improvement of social conditions and health safety of the countryside inhabitants. Absence of qualified physicians in the most of all countryside localities and presence of big distances to

Z. Hu et al. (Eds.): AIPE 2021, LNDECT 119, pp. 34–43, 2022.
https://doi.org/10.1007/978-3-030-97064-2_4

leading clinics situated in cities bring problems with timely medical care. Wide inculcation of telemedicine segment “patient – doctor” and Remote Health Monitoring (RHM) can help here. Now it is developed very quickly in various countries. Many research articles are published and many projects are implemented. Covid-19 pandemic has become an additional catalyst for RHM because face-to-face contacts of a patient and a doctor have become undesirable. So, RHM is very perspective now. There are many surveys, top-lists and research papers presenting practical projects and research works implemented in the world [1–4]. Usually the main attention is paid to technical problems connected with measuring various health parameters by sensors and medical devices and to transferring the data via network. Internet of Things (IoT) is quickly growing (including Medicine) segment of IT. There are also many works connected with specific medical problems solved with RHM help.

In our opinion, some shortage of investigations devoted to numerous subjective aspects of development and implementation of RHM takes place. Corresponding problems play there an important part and underestimation of them can spoil a good RHM project. In this article we try to accent concepts and methods uniting objective and subjective aspects, to display the problems and ways of their solutions.

A number of articles devoted to RHM in countryside and agriculture should be more. The survey [5] displays that the main accents in those articles are also connected with technical aspects. We think that RHM can improve much medical care for countryside. From other hand, RHM can be used not only for countryside inhabitants improving the social conditions but also for animals in stock-raising increasing effectiveness of agriculture. Certainly, some peculiarities of countryside and agriculture should be revealed and taken into account.

Usually some countryside inhabitant visits a doctor rarely. Under the RHM he does it some more seldom but a big number of the virtual contacts and corresponding generated medical data take place. The challenge is the shortage of qualified specialists (physicians for RHM for countryside inhabitants and veterinary surgeons for stock-raising). Artificial intelligence (AI) can help here taking a great part of the routine (uncreative) load. AI can make preliminary analysis of the information transferred from a patient and display the cases when personal attention of a doctor is not necessary. AI can also provide a patient and a doctor with relevant information, give to a patient typical answer to his typical questions, etc. These functions are implemented by RHM AI elements considered as Intelligent Agents (IA) [6, 7].

The main purposes of this paper are following.

1) To define possibility and expediency to use RHM for countryside and agriculture.
2) Output of the main development principles for the RHM systems with AI elements.
3) To discover or (and) create the RHM tools appropriate for countryside and agriculture.
4) Test of the elected tools.
5) To formulate the practical proposals for countryside and agriculture.

Our research background is based on scientific approaches and longitudinal investigations which were started in Medicine by A. Vorobiev [8]. A great attention is paid to a patient and to a doctor. Several general principles are followed from this view point.

- The medical care for a patient participating in monitoring must be better than without it ("not injure to a patient!").
- Life and work of a doctor participating in monitoring must be at least not worse than without it ("not injure to a doctor!").
- Artificial Intelligence shouldn't' exchange natural intelligence and professional skill of a physician but it should supplement them (AI can mean not only "Artificial Intelligence "but also "Augmented Intelligence").

A platform approach is used. The general basic telemedicine platform Medsenger. AI had been developed. Then various RHM systems for different categories of patients and classes of diseases had been created on a base of the platform. The results of a pilot exploitation of some of those systems are presented in this paper. The Medsenger.AI should be used in development of the RHM systems for countryside. The developed principles and methods should be used to create the "televeterinary" platform and various RHM systems for stock-raising. It can be also noted that our team had already created the veterinary platform "Petsenger" devoted to veterinary help to home animals. This experience should be also utilized.

2 Technical Principles and Structure of the Telemedicine Platform

The general principles are presented in the Introduction. Developed technical principles and structure of Medsenger.AI are following.

1) The telemedicine platform has general basic functions and convenient Application Program Interface (API) to integrate easy supplementing modules including ones created by outside developers. The key component of the platform is specific medical messenger "Medsenger" [9]. It organizes interaction between a patient and a doctor. In contradistinction to universal messengers (for example, WhatsApp), first of all, the Medsenger is connected with other subsystems of the platform. Therefore many important functions take place in the Medsenger which are absent in any universal messenger. Interaction between a patient and a doctor is implemented according with asynchronousprinciple: a patient asks when he has a question, a doctor answers when it is convenient for him but within preliminary agreed period.
2) All elements of interaction between a patient, a doctor and the system (in particular, the IA actions) are automatically recorded. Collected information is brought to form corresponding to Personal Electronic Medical Card (PEMC) and put in it [10]. A patient has a right of access to his PEMC and a right to open access for the doctors and officers of the medical institution. It is corresponding to the patient-centered concept and to Russian Federal Law about telemedicine. An attending doctor has to know his patient and a patient is able to choose his attending doctor. The subsystem belonging to the platform can form PEMC according with its algorithms. If the telemedicine system is applied in interaction with some already

existing PEMC in the Medical Information System of some medical institution that the outside interface of the platform can be used.

3) Presence of Decision Making Support Systems (DMSS) directed not only to a doctor but also to a patient. In our opinion, DMSS destined for a patient are needed and they have good perspectives. This IT and AI segment would be growing quickly soon. The main basic function of the DMSS in Medsenger.AI is automatic analysis of the information flow transferring from a patient to detect alarm situations when direct quick attracting attention of the physician is necessary. More complicated functions connected with more delicate estimation of the health statement can be added for the specific systems devoted to various categories of patients (for example, cancer patients, pregnant women, etc.) created on a base of the platform [11].
4) The platform includes the basic IA set. Other IA is added to the platform when some special RHM system is created on a base of the platform. Some of them can be created by the platform developers, other IA can be created by outside developers. There is a standard API for integration of all needed IA to the joint system. The various basic IA and their functions are described below.
5) During the RHM system running the choice of IA is done from preliminary developed wide IA set by the attending doctor for every patient individually. The doctor also accomplishes tuning of the chosen IA parameters according with the patient's category and health statement. If it is needed, the tuning can be done anew during monitoring period. The set of IA directly interacting with a patient in natural language can be assigned as a separate subsystem. Such IA implement "colloquial (or conversational) intelligence". They should help to a patient to generate questions and description of his health statement [12]. More details of IA are described in the next section.

3 Methods

The main method of AI implementation in the proposed platform is utilization of Intelligent Agents [13]. Some basic IA have simple behavior which implement standard determined algorithms necessary for all derived specific systems (for example, to remind about the time to take a drug or to measure temperature). Such IA is a software module developed without AI utilization. Other IA can process subjective conditions (for example, qualitative parameters of the patient health) and use AI approach. Various AI methods based on fuzzy sets and optimization can be implemented to analyze questionnaires filled up by patients [14, 15].

An important role is taken by IA implementing visualization of information. Various quantitative parameters (for example, blood pressure, temperature, etc.) and current events (for example, taking a drug) are presented as the graphs with one common time axis. The sample fragment can be seen in the Fig. 1.

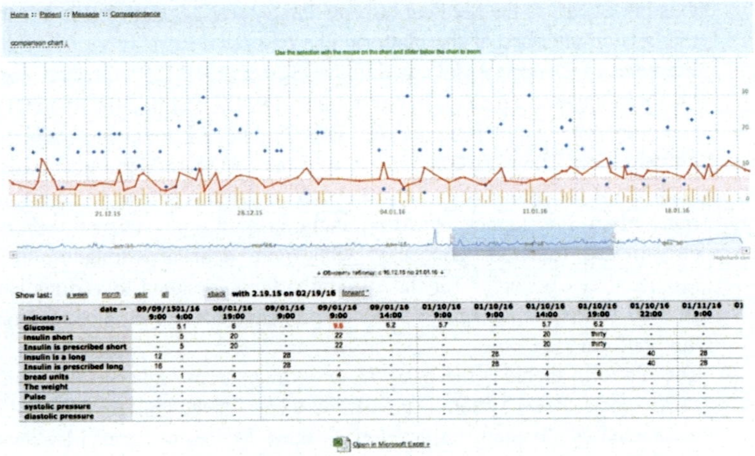

Fig. 1. Example of the graphical visualization

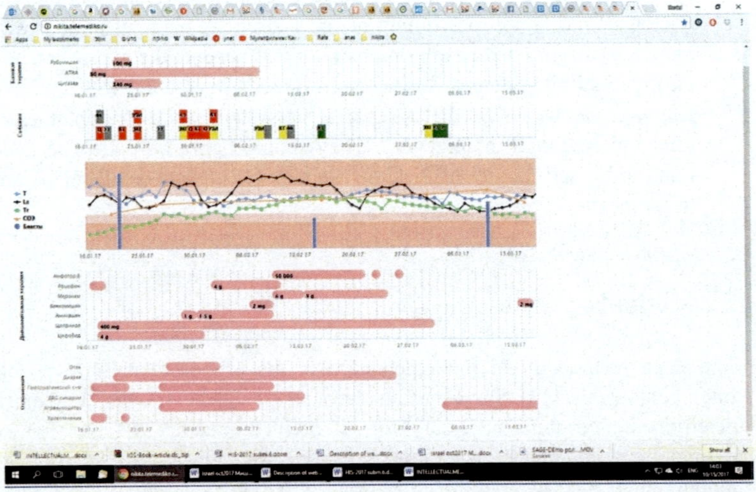

Fig. 2. An example of visualization utilizing the "Heat Map" together with the graphs

Specific "Heat maps" (HM) display the conditions on various symptoms and appreciation of the health statement in the whole taking into account the "weight" of different symptoms (including quantitative and qualitative). The sample of the HM joined with the graphs is presented in the Fig. 2.

United "Heat Maps" are used for a group of patients regarding one attending physician. It is convenient for the doctor when he has many patients under RHM.

IA can have high level of artificial intelligence utilization for more delicate estimation of the symptoms and health statement with combination of quantitative and qualitative parameters including nominating weight factors in the integrated parameter of a patient health. The fuzzy sets approach can be also used here [16, 17].

Conversational IA are utilized for support of the chat interaction between the system and a patient. They can also have different level of "intelligence": from simple chat bots enumerating questions from preliminary set demanding standard answers (for example, yes/no, besides, sending standard questionnaires to a patient, etc.) to IA which can help a patient to formulate his questions to the doctor and generate typical answers even to non-standard questions [18].

4 Results of the Telemedicine Platform Pilot Running

Several RHM systems had been already created on a base of the telemedicine platform Medsenger.AI: ONCONET and ONCO REHUB for cancer patients after hard treatment, TRANSPLANTNET for patients after organ transplantation, Petsenger for veterinary help to home animals, COVID REHUB for rehabilitation of patients after COVID-19 illness, etc. All created systems had been tested and the pilot exploitation took place. Below some results are presented. The system ONCONET was used during 2018–2020 in 22 medical organizations in 10 regions of Russia. 174 doctors and 382 patients participated in the pilot running the system [19]. Similarity in composition of diagnoses, stages of diseases and treatment regimens, as well as on the sex-age composition was observed.

Comparison of quantity (frequency) of complications which took place after hard treatment (chemotherapy) between cancer patient group participating in monitoring and the reference group had been done. The main results are seen in the Table 1.

Table 1. Quantity (frequency) of complications

Complications type	Group under RHM,%	Reference group,%	Statistical significance of the difference, YES/NO
Hematological	32.5	41.6	YES
Nephrotoxic	3.6	13.6	YES
Gastrointestinal	0.7	2.6	YES
Hepatotoxic	29.1	23.0	NO
Neurotoxic	0	3.2	YES
Skin	0	1.6	YES
Total	65	78	YES

Besides, the postponements of the next chemotherapy course took place significantly less under monitoring (7.9% for ONCONET vs. 11.9% in the reference group). It is an important parameter characterizing total influence of the complications on the rehabilitation progress and affecting further health of the patient.

Statistical significance of the differences noted in the Table 1 was estimated according with formula of criterion Q for comparison of binary samples [20]:

$$Q = \frac{p_1^* - p_2^*}{\sqrt{\frac{p_1^*(1-p_1^*)}{n_1} + \frac{p_2^*(1-p_2^*)}{n_2}}}$$

Here p_1^* and p_2^* are the "Yes appearance" frequencies for binary samples 1 and 2 correspondingly; n_1 и n_2 – numbers of patients in the samples. "Yes appearance" means here a fact that the complication took place for the concrete patient. The critical Q value was taken as 1.96. It is corresponding to error (to reject the true hypothesis) probability 0.05. Presence of "YES" in the fourth column in the Table 1 means that $Q > 1.96$ and the difference is significant. Presence of "NO" means that $|Q| < 1.96$ and the difference is not significant. Various supplementing practical nuances of similar implementation of binary samples comparison were described in [21].

The RHM system COVID REHAB had been created on a base of the platform Medsenger.AI after COVID-19 pandemic start. The system is intended for the remote rehabilitation of patients who had earlier the COVID-19 illness. 1202 patients (till April 2021) took part in the pilot running of the RHM system organized in the National Medical Research Center of Rehabilitation and Health Resort (Federal institution situated in Moscow) and 830 patients took part being in connection with medical institutions situated in 5 regions of Russia. Some part of the patients included countryside inhabitants and there were countryside localities from which only one person participated in the project. The wide set of various results deserves to be described in the separate article. Here it can be noted that complications cases number in the group taking part in the pilot project was reduced approximately on 20% as compare with the reference group.

The most of all doctors and patients estimated their participation in the pilot RHM projects as positive.

So, the objective and subjective results of the pilot utilization of the proposed systems were obviously good for the patients and for the doctors too.

5 Discussion

The above results confirm that proposed telemedicine platform allows create RHM systems being effective and convenient tools to provide remote health monitoring. The treatment and rehabilitation medical results received in the focus group of patients participating in RHM were obviously better than in the reference group of patients without RHM implementation. Involved doctors and patients estimated positively utilization of RHM and the proposed systems.

We hope that some more other similar RHM systems will be developed on a base of the proposed platform in the nearest future. All of them can be used for patients living in countryside, concerning relevant category of patients and class of diseases. Probably it is expedient to develop the specific RHM system for countryside inhabitants which haven't yet significant problems with health. We hope that they will save their health and RHM will help them to do it. So, the RHM system for healthy people living in countryside and wanting to save their health should be created. Some peculiarities typical for them should be taken into account. Preliminary research devoted to exposure of those peculiarities should be done together with doctors, representatives of the countryside community or (and) with managers working in agriculture. Certainly, the peculiarities take place but it should be needed to amplify them.

The proposed telemedicine platform should be developed further. Besides of direct improvement of medical care, an opportunity to get information support for research is also important. Set of the PEMC should be a great information resource. It can be analyzed to reveal important regularities concerning big groups of patients. The data accumulated in the PEMC have to be depersonalized and consolidated. Data Mining approach should be used and AI should take important role there because the data would have peculiarities typical for the AI field: not enough definite, not enough complete to be processed by algorithm software. A combination of precise and fuzzy parameters would take place and flexible approach combining clear algorithms and AI methods would be needed. To implement practical utilization of the revealed regularities Machine Learning (ML) can be used in RHM DMSS in combination with Expert approache [22]. It can be also noted that RHM allows make analysis of the temporal medical data collected for a concrete patient. It opens wide opportunities for real personalization of medical care. Research works devoted to development of corresponding methods should be implemented [23, 24].

6 Conclusion

The main purposes of this paper displayed in the Introduction have been reached.

1) RHM can be used for countryside and agriculture. The telemedicine platform Medsenger.AI and developed on its base RHM systems for various patients categories and classes of diseases (ONCONET and ONCO REHAB for cancer patients, TRANSPLANTNET for patients after organ transplantation, COVID REHAB and others) have been tested successfully. The quantitative objective results were good as well as the subjective estimation by doctors and patients who was participating in the test running. During the pilot exploitation of ONCONET and COVID REHAB some number of countryside inhabitants took part in those projects and the participation was successful. So, those systems can be already now used for countryside. It is expedient to involve people living in country side to RHM more.
2) Developed theoretical principles of the RHM systems creation including utilization of IA sets would be important and useful for development of new RHM systems. In particular, the specific RHM system for health safety of countryside inhabitants can be created on a base of the platform Medsenger.AI.
3) The similar approach, developed theoretical principles and accumulated experience can be used to develop the platform for remote veterinary monitoring in stock-raising. It is expedient but it demands to conduct multidisciplinary research with participation of IT and AI specialists together with veterinary surgeons. Such

research and implementation of the RHM approach would be useful much for agriculture.

4) Further development of the proposed telemedicine platform and the RHM systems should be associated with an expansion of AI functions. The opportunities of intelligent analysis of the collected and depersonalized medical data should be realized. It would allow bring out important general regularities. From other hand, longitudinal remote health monitoring would generate the big temporal data sets corresponding to every patient. Intelligent analysis of such data sets is expedient and perspective. Besides of these two ways connected with DMSS, further development of conversational IA should also take place. Certainly, the extensive and intensive research work is necessary to support all these ways.

Acknowledgments. This work was supported by the Ministry of Science and Higher Education of the Russian Federation (AAAA-A18-118012390247-0), by RFBR grant (project № 19-07-01235).

References

1. Moreno, H.B.R., Ramírez, M.R., Hurtado, C., Lobato, B.Y.M.: IoT in medical context: applications, diagnostics, and health care. In: Chen, Y.-W., Zimmermann, A., Howlett, R.J., Jain, L.C. (eds.) Innovation in Medicine and Healthcare Systems, and Multimedia. SIST, vol. 145, pp. 253–259. Springer, Singapore (2019). https://doi.org/10.1007/978-981-13-8566-7_25
2. http://evercare.ru/remote-monitoring-top10. Accessed June 2021
3. Zingerman, B.V., Vorobyev, A.I., Shklovsky-Kordy, N.E.: About telemedicine: patient-doctor. Phys. Inf. Technol. **1**, 61–79 (2017)
4. Parthiban, N., et al.: Remote monitoring of implantable cardioverter defibrillators: a systematic review and meta-analysis of clinical outcomes. J. Am. Coll. Cardiol. **65**, 2591–2600 (2015)
5. Palanisamy, S., Selvaraj, R., Ramesh, T., Ponnusamy, J.: Applications of remote sensing in agriculture - a review. Int. J. Curr. Microbiol. App. Sci. **8**(01), 2270–2283 (2019). https://doi.org/10.20546/ijcmas.2019.801.238
6. Rosenfeld, A., Kraus, S.: Strategical argumentative agent for human persuasion. In: ECAI, vol. 16, pp. 320–329 (2016)
7. Isern, D., Moreno, A.: A systematic literature review of agents applied in healthcare. J. Med. Syst. **40**(2), 43 (2016). https://doi.org/10.1007/s10916-015-0376-2
8. Shklovskiy-Kordi, N., Shifrin, M., Zingerman, B., Vorobiev, A.: Some directions of medical informatics. In: Siuly, S., et al. (eds.) Health Information Science, vol. 10594, pp. 40–44. Springer, Cham (2017). https://doi.org/10.1007/978-3-319-69182-4_3 ISSN: 03029743 978-331969181-7
9. Shklovskiy-Kordi, N., Borodin, R., Zingerman, B., Shifrin, M., Kremenetskaya, O., Vorobiev, A.: Web-service medical messenger – intelligent algorithm of remote counseling. In: Siuly, S., Lee, I., Huang, Z., Zhou, R., Wang, H., Xiang, W. (eds.) Health Information Science, vol. 11148, pp. 193–197. Springer, Cham (2018). https://doi.org/10.1007/978-3-030-01078-2_18
10. Zingerman, B.V., Shklovsky-Kordi, N.E.: Electronic medical record and the principles of its organization. Phys. Inf. Technol. **2**, 37–57 (2013)

11. Zingerman, B.V., Shklovsky-Kordy, N.E., Karp, V.P., Vorobyev, A.I.: Integrated electronic medical card: tasks and problems. Phys. Inf. Technol. **1**, 24–34 (2015)
12. Galitskiy, B.: Chatbots for CRM and dialogue management. In: Artificial Intelligence for Customer Relationship Management, pp. 1–61. Springer, Cham (2021). https://doi.org/10.1007/978-3-030-61641-0_1
13. Shinkariov, S., et al.: Telemedicine system with elements of artificial intelligence for health monitoring during COVID-19 pandemic. In: Huang, Z., Siuly, S., Wang, H., Zhou, R., Zhang, Y. (eds.) HIS 2020. LNCS, vol. 12435, pp. 103–110. Springer, Cham (2020). https://doi.org/10.1007/978-3-030-61951-0_10
14. Evelson, L.I., Dubovoy, I.I., Borisova, E.P.: Improving screening methods to determine belonging to the risk group for the consequences of alcohol effect (with information technology and mathematical modeling). Phys. Inf. Technol. **3**, 6–17 (2018)
15. Alguliyev, R., Imamverdiyev, Y., Suchostat, L.: Anomaly detection based on optimization. Int. J. Intell. Syst. Appl. (IJISA) **9**(12), 87–96 (2017). https://doi.org/10.5815/ijisa.2017.12.08
16. Tripathy, B.K., Agrawal, A., Reddy, A.J.: A Comparative Analysis of firefly and fuzzyfirefly based kernelized hybrid c-means algorithms. Int. J. Intell. Syst. Appl. (IJISA) **11**(6), 49–68 (2019). https://doi.org/10.5815/ijisa.2019.06.05
17. Dubovoy, I.I., Evelson, L.I., Borisova, E.P., Green, M.S., Taratushkina, O.A., Mitina, O.V.: Application of information technologies for the selection of person, who has some problems connected with alcohol or risk of such problems. Prophyl. Med. **20**(1(2)), 24–25 (2017)
18. Galitskiy, B.: Adjusting chatbot conversation to user personality and mood. In: Artificial Intelligence for Customer Relationship Management, pp. 93–127. Springer, Cham (2021). https://doi.org/10.1007/978-3-030-61641-0_3
19. Kargalskaya, I.G., Shinkarev, S.A., Zingerman, B.V., Fistul, I.A., Nozik, A.V., Arseniev, S. B.: IADIS International Conference e-Health Virtual Conference, pp. 193–198. ISBN: 978-989-8704-18-4
20. Orlov, A.: Applied Statistics, p. 671 (2006). ISBN: 5-472-1 122-1
21. Geger, E.V., Kozlova, I.R., Yurkova, O.N., Evelson, L.I.: Methodology for comparing binary samples in the analysis of medical data for making managerial decisions. XXI century: results of the past and problems of the present plus. Comput. Sci. Comput. Eng. Manag. **9**(2(50)), 164–170 (2020)
22. Goldberg, S., Pinsky, E., Galitsky, B.: A bi-directional adversarial explainability for decision support. Hum.-Intell. Syst. Integr. **3**(1), 1–14 (2021). https://doi.org/10.1007/s42454-021-00031-5
23. Pathak, A.R., Pandey, M., Rautaray, S.: Adaptive model for dynamic and temporal topic modeling from big data using deep learning architecture. Int. J. Intell. Syst. Appl. (IJISA) **11**(6), 13–27 (2019). https://doi.org/10.5815/ijisa.2019.06.02
24. Hatsek, A., Shahar, Y., Taieb-Maimon, M., Shalom, E., Klimov, D., Lunenfeld, E.: A scalable architecture for incremental specification and maintenance of procedural and declarative clinical decision-support knowledge. Open Med. Inform. J. **4**, 255–277 (2010)

Neuroeconomic Modeling of Distributed Barter Systems

I. V. Stepanyan[1(✉)] and M. A. Chirkov[2]

[1] Mechanical Engineering Research Institute of the Russian Academy of Sciences (IMASH RAN), 4 Maly Kharitonyevsky Pereulok, Moscow 119991, Russia

[2] Lomonosov Moscow State University, Moscow 119991, Russia

Abstract. The purpose of this work is to analyze the capabilities of automated barter chains based on decentralized technologies - distributed databases and peer-to-peer networks. Blockchain technologies are taken as an example. The authors describe the concept of the so-called game, which is a system-forming platform of automated barter chains for effective interaction between participants in economic processes (players) and aimed at the development and acquisition of wealth of each participant, as well as the disclosure of his creative and personal potential. As a result of the analysis, the function of the equivalent in money-free exchange is determined and it is concluded that it is possible to create a distributed mentoring institute, where every realized talent will take on apprentices and be responsible for their development. This article provides the analysis of questions of generalization of the financial paradigm in the transition to a new socio-economic technological order by changing cultural attitudes to the model of effective, targeted use of resources in the amount necessary for the disclosure of the creative potential of each player and in the amount necessary for his personal sense of happiness and harmony with himself, the surrounding society and nature (symbiotic, ecological, systematic approach). The presented concept is proposed as a basis for ensuring sustainable development of society by redirecting the resource base to the harmonious sustainable development and interaction of man and nature, the formation of an ecocentric social consciousness that determines the ethical attitude of man to The Earth's biosphere, to flora and fauna.

Keywords: Distributed systems · Blockchain · Sustainable development · Neuroeconomic modeling · Economic relations · Barter software

1 Instruction

The development of information systems is largely associated with the use of nature-like technologies and the development of distributed software and hardware solutions. This can be seen in neural network technologies, where instead of a single central processor, a set of elementary calculators — neurons is used. This also applies to the emergence of decentralized algorithms and databases: distributed hash tables (DHTs), interplanetary file system (IPFS), various mechanisms (Filecoin, Bitswap, BigchainDB, Ethereum, Swarm), MaidSafe technology, the purpose of which is to provide

Z. Hu et al. (Eds.): AIPE 2021, LNDECT 119, pp. 44–51, 2022.
https://doi.org/10.1007/978-3-030-97064-2_5

decentralized communication throughout the world, and others [1–9]. In peer-to-peer (p2p) networks, all networked computers have identical client and server functions [10, 11]. All these systems have increased reliability of transactions, calculations, and other technological processes due to the properties of decentralization and powerful encryption algorithms. This circumstance leads to a rethinking of the economic models established around the world [12-15]. Consider a retrospective review of economic systems and the development of monetary relations based on decentralized networks.

Barter in prehistoric times was considered as a sporadic exchange of products or services. With the growth of labor productivity and the commodity mode of production, there was a need for a universal equivalent, which led to the appearance of money. In a developed capitalist economy, barter occupies a marginal niche, being used, for example, in conditions of lack of liquidity. Money is performed including the role of a value measure and a means of circulation, allowing you to divide in time the moments of alienation by a participant in the turnover of goods a and the acquisition of goods B, C, etc., the total cost of which is equal to the cost of goods A. Early views of economists, philosophers [16] on this issue are associated with the lack of technological prerequisites at that time.

The development of peer-to-peer networks, technologies based on encryption and distributed databases has become the basis of a number of new economic models. It is obvious that the function of money has begun its transformation and this process is difficult to regulate due to technological progress. This is confirmed by the intensive development of various cryptocurrencies.

One of the main technologies is the blockchain, which has proven itself as a tool for economic activity and a tool for distributed information systems (similar to IOTA technology. Technically, it is a distributed crypto database, which is a chain of «blocks» stored simultaneously on multiple computers - a growing list of records that are linked using cryptography. Each such block contains the cryptographic hash of the previous block, a timestamp, and transaction data.

The Open Bazaar project is a blockchain-based trading platform. In general, it is similar to the discussed tools, here it is not barter, but buying and selling for cryptocurrency. There are also known crowdsourcing platforms based on blockchain [9] and "people's court" systems based on blockchain – crowdjury.org.

A useful feature of this technology is the ability to capture information and thus avoid misinterpretations of agreements (guarantee the reliability of the agreement and the indisputability of transactions). This is an analog of a public digital signature, which cannot be canceled, because it is stored simultaneously on all devices participating in the general economic activity.

This paper presents the concept of a recommendation system for managing barter chains based on decentralized crypto databases and peer-to-peer networks (hereinafter referred to as the game). The concept is based on the following concepts:

* «Desires» (a goal that allows the player to improve the quality of life, designed and published as a message of need);

* «Creation» (creative physical and mental activity of the player that brings him satisfaction and is aimed at improving the game results and benefits the players);

* «Affluence» (having decent housing, freedom and means of movement, a stable balanced diet, physical and mental health).

2 Materials and Methods

Consider the basic principles of interaction based on automated barter chains. The game is an automated platform for the direct exchange of values, including goods and services with a direct targeting system (P2P-Marketplace). It is a decentralized system of consolidation and mutual assistance; it has a simple interface and is accessible from electronic devices. Allows you to build self-organizing communities aimed at the development, interaction and prosperity of the communities themselves. It has an interface similar to a Bulletin Board with catalogs of products and services published by users, without any restrictions on their personal pages, which are an evolution of the card with information about the user.

The automated search for barter chains is based on an abstract scheme of a network decentralized Game management system - a network graph, each vertex of which-a participant in the game (a living person, a subject) has two lists: a wish list and a list of opportunities (including skills). The general algorithm of the "Game" is that the subjects (hereinafter referred to as players, avatars) agree on barter transactions in the formalism of the Game: the blockchain records the fact of an agreement on a barter transaction and its completion. Trading operations can be delegated to one or more avatars for management. To optimize economic activity and logistics, various optimization algorithms are provided for building rational barter chains in the interests of participants. You can use internal tokens to transfer value or smart contracts [17–19].

The proposed system of smart contracts, organized on the acyclic directed graph principle for Game implementation, is suitable for flexible management of planning, design, progress, time reserves, reporting, private asset management, process support, industrial production, and structuring complex processes in decentralized Autonomous organizations. Targeting in such a system is based primarily on a direct sales model and advertising that is interesting to end users (allowed by them to display).

By joining the Game, a new participant contributes a resource to the game society as a share contribution in accordance with the law on consumer cooperation [20–22]. Resource - a unit or set of intellectual or physical activities. When entering the game, a person notifies the system about what, in what quantity and in what time period he can add to the game fund, for example: products, things [23], services, training, development, any useful and/or demanded actions. In this case, the share payment is entered into his personal account and is his property. The player has the right to dispose the unit fee independently and manage the services of other players, as well as top up the unit fee of the Deposit and confirmation of qualification.

Resources are distributed among players based on the principle of "resources for creative potential" to build this potential into a common collective playing field. The priority is to allocate resources that will directly strengthen the territory of the agreements and the game system as a whole. For this purpose, each user has a personal rating system and a system of social groups, on the basis of which various privacy settings can be applied and certain functions can be enabled. It allows you to adjust flexibly the filtering and security and to provide a comfortable social interaction. Private resources are resources that represent a manifestation of work or creativity. The Union has the right to transfer part of its experience to a new participant who has joined the Game, subject to the consent of all members of the Union.

Thus, the game implementation technologies are represented as a convergence of historical, cultural and scientific heritage, as well as technologies in the field of telecommunications, marketing, distributed registries, peer-to-peer communication, queuing, machine learning and social engineering.

For the neuroeconomic modeling of the Game, we use a graph-theoretic approach: we associate a participant in economic relations (player) with each vertex of the graph, and the presence of economic interaction with the arcs. The simplest variant of interaction is that two players enter into economic relations, while the first player supplies the resource to the second, and the second, instead of monetary settlement with the first, uses a chain of economic interactions, which is closed to the first player. Thus, the barter chain functions, ensuring that the needs of the first and second players are met.

Obviously, within the formalism of the proposed model, the number of feedbacks in the form of barter chains will determine the level of economic stability of the entire system. In graph theory, such feedbacks are simple cycles.

For further reasoning, we will create a random growing graph and estimate the probability of occurrence of simple cycles in it that simulate economic feedbacks. Consider the algorithm:

1. A set of N nodes (vertices) N = {1,2,3 .... N} is given. Each node is numbered.
2. The initial structure of the graph is set from two connected nodes.
3. Repeat steps 4–7 N times.
4. The structure of the graph is recalculated
5. One of the N nodes is chosen at random and connected to an arbitrary node in the graph
6. The number N of free vertices decreases by one
7. Go to step 3.
8. Stop.

3 Results

For model experiments and estimating the scatter of the statistical values of the parameters of growing graphs, this algorithm was repeated a hundred times. After each run of the algorithm, the parameters were calculated and curves were plotted. These curves were then superimposed on each other to visualize the final results (Fig. 1).

The author's hypothesis is that the presence of multiple feedbacks makes any (including economic) system more stable.

Similar effects with feedback are observed in the neural network structure of the brain of higher animals and humans. In this sense, distributed networking technologies have biological counterparts, proven by natural selection and evolution.

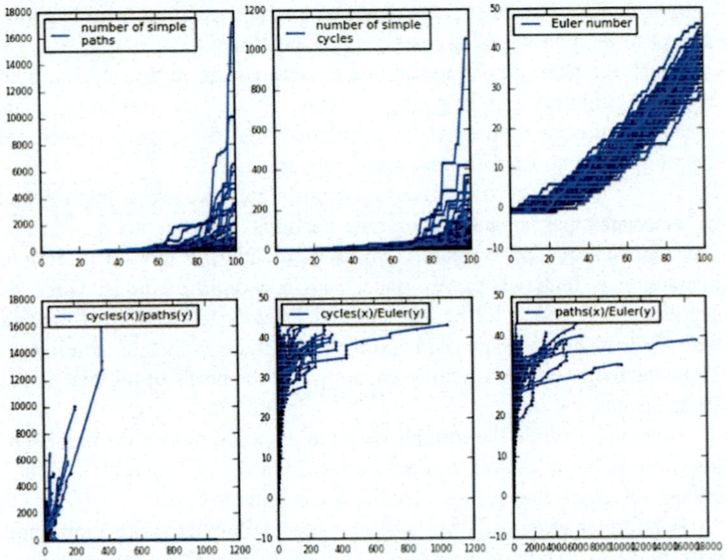

Fig. 1. 100 iterations of graph growth (100 experiments). Top: abscissa - algorithm step, ordinate - parameter value. Bottom: abscissa - the number of simple cycles, ordinate - the number of simple paths (left); the abscissa is the number of simple cycles, the ordinate is Euler's number (middle); the abscissa shows the number of simple paths, and the ordinate shows Euler's number (on the right).

Despite the simplicity of the algorithm, interesting features are observed in the structure of the graph: the probability of a sharp increase in cyclic structures (by 5 and 7 times) is observed in 2% of cases starting from 83 growth steps.

Consider an example of economic efficiency and comparison with the free market model. The disadvantages of the modern economic model are clearly visible in the following typical example. At the beginning of economic activity, one need to take a loan, then spend money on advertising, marketing and searching for counterparties, then return the loan in conditions of inflation and systemic crises.

On the other hand, within the framework of the Game concept, the player forms a desire, and a unique identifier is assigned to the desire. From this moment, the search mechanism aimed at meeting the needs and desires of the maximum number of participants of the barter chain is launched. When the necessary barter chain is found, the cost of the materials and transport delivery (decentralized, national logistics) is formed with fixing all the necessary parameters on the blockchain.

Upon delivery of the product to the customer, the integrity of the packaging and goods is recorded, and compliance with the declared quality and terms of the smart contract is checked, with the possibility of instant confirmation by online observers, if everything went well, the smart contract distributes points and fixes the end of the cycle.

The self-organization of the game allows to optimize not only the economic component, but also the environmental (non-waste) production principles [24–26]. This is a redirection of the resource base to the harmonious more sustainable development of man and nature, the formation of an ecocentric social consciousness, which determines the ethical attitude of man to the Earth's biosphere, flora and fauna [27]. Ecocentric thinking suggests that the biosphere, flora and fauna are not a utilitarian application to humans, but equal to humanity.

4 Discussion

It should be noted that the goal aimed at obtaining wealth (the concept of the game) does not coincide with the goal of making profit (the concept of a common economic model). These are totally different paradigms. Moreover, the proposed model can be used to make a profit, since it is more general.

The question arises: how can the equivalent function be exchanged without money? Tokens (points and their various types - ratings of smart contracts) can have this function, for example, this can be a number of positive reviews. Of course, it is important to use the correct established contract form (smart contract) [28].

The formation of exchange baskets and transaction conditions should be formalized very clearly and in detail in accordance with the legislation of the country (unless otherwise specified in the contract, then it is considered that the goods are considered equivalent - the contract should be drawn up so that it really should not be otherwise). In the Russian Federation, this is Article 568 of the Civil Code - a barter agreement [29].

The relevance and prospects of this topic are due to the growth in the speed of cryptocurrency transactions, which currently reach 50,000 transactions per second (Table 1), as well as a number of publications on this topic [for example, [30–32].

Table 1. Top 5 transaction speed cryptocurrencies in 2021

Cryptocurrency	Transactions/second
Solana (SOL)	50000
Ripple (XPR)	1500
Celo (CELO)	1000
Algorand (ALGO)	1000
Cardano (ADA)	257

5 Conclusions

The article demonstrates the paradox of double negation in the historical development of circulation: direct exchange (barter) - exchange mediated by the universal (monetary) equivalent (commodity - money - commodity formula) - barter in the information society (smart contracts and barter chains without money).

The software platform within the framework of the described concept allows achieving a high level of automation with maximum reliability and guaranteeing the security of transactions.

Software switching and network virtualization technologies with peer-to-peer connections between users, devices and applications allow implementing the proposed decentralized microservice architecture using open source. This will contribute to the sustainable development of society and its harmony with nature.

References

1. Crosby, M., et al.: Blockchain technology: beyond bitcoin. Appl. Innov. **2**(6–10), 71 (2016)
2. Shrier, D., Wu, W., Pentland, A.: Blockchain & infrastructure (identity, data security). Massachusetts Inst. Technol.-Connection Sci. **1**(3), 1–19 (2016)
3. Sankar, L.S., Sindhu, M., Sethumadhavan, M.: Survey of consensus protocols on blockchain applications. In: 2017 4th International Conference on Advanced Computing and Communication Systems (ICACCS). IEEE (2017)
4. Sharma, P.K., Moon, S.Y., Park, J.H.: Block-VN: a distributed blockchain based vehicular network architecture in smart city. J. Inf. Process. Syst. **13**(1), 184–195 (2017)
5. Bashir, I.: Mastering Blockchain: Distributed Ledger Technology, Decentralization, and Smart Contracts Explained. Packt Publishing Ltd, Birmingham (2018)
6. Panescu, A.T., Vasile, M.: Smart contracts for research data rights management over the ethereum blockchain network. Sci. Technol. Libr. **37**(3), 235–245 (2018)
7. Swan, M.: Blockchain for business: next-generation enterprise artificial intelligence systems. Adv. Comput. **111**, 121–162 (2018)
8. Zibin, Z., et al.: Blockchain challenges and opportunities: a survey. Int. J. Web Grid Serv. **14** (4), 352–375 (2018)
9. Kogias, D.G., et al.: Toward a blockchain-enabled crowdsourcing platform. IT Prof. **21**(5), 18–25 (2019)
10. El-Ansary, S., Alima, L.O., Brand, P., Haridi, S.: Efficient broadcast in structured p2p networks. In: Kaashoek, M.F., Stoica, I. (eds.) Peer-to-Peer Systems II, vol. 2735, pp. 304-314. Springer, Heidelberg (2003). https://doi.org/10.1007/978-3-540-45172-3_28
11. Kamvar, S.D., Schlosser, M.T., Garcia-Molina, H.: The eigentrust algorithm for reputation management in p2p networks. In: Proceedings of the 12th International Conference on World Wide Web (2003)
12. Lun, D.S., et al.: Achieving minimum-cost multicast: a decentralized approach based on network coding. In: Proceedings IEEE 24th Annual Joint Conference of the IEEE Computer and Communications Societies, vol. 3. IEEE (2005)
13. Kaur, G., Sharma, R.D.: Voyage of marketing thought from a barter system to a customer centric one. Mark. Intell. Plann. (2009)
14. Elsayed, W.T., El-Saadany, E.F.: A fully decentralized approach for solving the economic dispatch problem. IEEE Trans. Power Syst. **30**(4), 2179–2189 (2014)
15. Yukun, L., Tsyvinski, A.: Risks and returns of cryptocurrency. Rev. Financ. Stud. **34**(6), 2689–2727 (2021)
16. Marx, K.: Das kapital: kritik der politischen ökonomie; vol. 1. Dietz Verlag (1977)
17. Luu, L., et al.: Making smart contracts smarter. In: Proceedings of the 2016 ACM SIGSAC Conference on Computer and Communications Security (2016)
18. Goel, S., et al.: ZEUS: analyzing safety of smart contracts (2018)

19. Tsankov, P., et al.: Securify: practical security analysis of smart contracts. In: Proceedings of the 2018 ACM SIGSAC Conference on Computer and Communications Security (2018)
20. Flood, J.: The collegial phenomenon: the social mechanisms of cooperation among peers in a corporate law partnership by Emmanuel Lazega, pp. 291–294 (2005)
21. Bulnina, I.S., et al.: Public and private partnership as a mechanism of government and business cooperation. Mediterr. J. Soc. Sci. **6**(1 S3), 453–453 (2015)
22. Raustiala, K.: The architecture of international cooperation: transgovernmental networks and the future of international law. Va. J. Int'l L. **43**, 1 (2002)
23. Skarmeta, A.F., Hernandez-Ramos, J.L., Victoria Moreno, M.: A decentralized approach for security and privacy challenges in the internet of things. In: 2014 IEEE World Forum on Internet of Things (WF-IoT). IEEE (2014)
24. Alexander, G.: An eco-centric approach to sustainable community development. Community Dev. J. **41**(1), 104–108 (2006)
25. Borland, H., Lindgreen, A.: Sustainability, epistemology, ecocentric business, and marketing strategy: ideology, reality, and vision. J. Bus. Ethics **117**(1), 173–187 (2013). https://doi.org/10.1007/s10551-012-1519-8
26. Dudin, M., et al.: Methodological approaches to classification of innovation potential in the context of steady development of entrepreneurial structures. World Appl. Sci. J. **27**(13A), 563–566 (2013)
27. Palla, G., et al.: Uncovering the overlapping community structure of complex networks in nature and society. Nature **435**(7043), 814–818 (2005)
28. Abdullah, R.S., Faizal, M.A.: Block chain: cryptographic method in fourth industrial revolution. Int. J. Comput. Netw. Inf. Secur. (IJCNIS) **10**(11), 9–17 (2018). https://doi.org/10.5815/ijcnis.2018.11.02
29. Andrei, Y.: Barter in the Russian economy: classifications and implications (evidence from case study analyses). Post-Communist Econ. **12**(3), 279–291 (2000)
30. Wang, L., Wang, G.: Big data in cyber-physical systems, digital manufacturing and industry. Int. J. Eng. Manuf. (IJEM) **6**(4), 1–8 (2016). https://doi.org/10.5815/ijem.2016.04.01
31. Kuznetsov, A., Oleshko, I., Tymchenko, V., Lisitsky, K., Rodinko, M., Kolhatin, A.: Performance analysis of cryptographic hash functions suitable for use in blockchain. Int. J. Comput. Netw. Inf. Secur. (IJCNIS) **13**(2), 1–15 (2021). https://doi.org/10.5815/ijcnis.2021.02.01
32. Akter, O., Akther, A., Ashraf, U., Manowarul, I.: Cloud forensics: challenges and blockchain based solutions. Int. J. Wirel. Microw. Technol. (IJWMT) **10**(5), 1–12 (2020). https://doi.org/10.5815/ijwmt.2020.05.01

The Use of Vegetation Indices in Comparison to Traditional Methods for Assessing Overwintering of Grain Crops in the Breeding Process

Rashid Kurbanov[1], Natalia Zakharova[1(✉)], Vladimir Sidorenko[2], and Sergey Vilyunov[2]

[1] Federal Scientific Agroengineering Center VIM, Moscow, Russian Federation
smedia@vim.ru
[2] Federal Scientific Center of Legumes and Groat Crops, Oryol, Russian Federation

Abstract. A breeder's job is a creative process that requires the routine collection and painstaking personal analysis of many types of data. This information is often subjective and cannot be statistically processed, relying only on the intuition of a scientist. One of these criteria for evaluating the source material is the visual assessment of overwintering varieties and breeding lines of winter crops. Overwintering expressed in points is approximate, and it is impossible to reliably compare breeding lines using it, especially when the points are approximately equal. For the first time in Russia, in the fields of FSBSI "Federal Scientific Center of Legumes and Groat Crops" (FSBSI FSC LGC) located in the Oryol region, elements of the developed method of remote sensing of small areas in the breeding process were tested based on "winter hardiness" of winter wheat, reliably expressed by vegetation indices. To assess breeding lines and crop varieties for winter hardiness, data were collected using unmanned vehicles at an altitude of 50 m and multispectral cameras with a resolution of 2 cm/pix. of Federal Scientific Agroengineering Center VIM. Previously, to determine the date of photography, an analysis of the weather conditions of the region for 20 years was carried out. It was revealed that regardless of the date of snow melting, the time of resumption of the spring vegetation of winter crops is located around the date of April 12. Accordingly, the third decade of April is recommended for optimal remote sensing of awakening crops in the region. Comparison of the data of vegetation indices (NDVI, NDRE, ClGreen) with the traditional results of visual scoring showed high reliability of the obtained digital data (correlation coefficient $r > 0.7$) for all three indices. It was noted that the NDVI index is more informative since it carries additional information not only about the preservation but also about the diversity of crops, i.e. on the presence of areas with fully or partially fallen out vegetation. The focal sparseness of crops was maximally characterized by the standard deviation of the NDVI index, increasing by 10…15% in the "dropped out" areas (correlation coefficient $r = -0.55$) in comparison to the other two. An objective statistical assessment of digital information from small areas (plot area of 8 sq. m.) on 29 variants made it possible to additionally identify 6 cultivars (with a score of overwintering, comparable to the reference cultivar for the region) and compare the 18

Z. Hu et al. (Eds.): AIPE 2021, LNDECT 119, pp. 52–64, 2022.
https://doi.org/10.1007/978-3-030-97064-2_6

variants that stood out among them. This would be impossible with traditional visual methods. In general, remote assessment of winter crops overwintering using vegetation indices based on comparison to an adapted reference variety allows you to quickly and without the involvement of a highly specialized and highly qualified specialist in the field of plant growing accumulate objective information from year to year on the overwintering of a characteristic collection of winter crops. Accumulation of such objective structured digital information for different years, tied to the data of the reference variety, will allow the formation of data arrays for further machine learning of specialized neural networks.

Keywords: Winter wheat · Unmanned aerial vehicles (UAV) · Breeding · Aerial data · Vegetation indices · Big data

1 Introduction

In agricultural production, the use of technology for the optical assessment of the vegetation cover of field crops is expanding [1, 2]. The vegetation index (NDVI) obtained from hand sensors and satellites is widely used [3], which is a convenient digital tool for analyzing the quality of development of individual crops, including winter wheat [4, 5]. Transition to remote optical sensing is a promising scientific and technical project in the field of mechanization and robotization of agriculture [6, 7]. In the breeding process of winter wheat, in addition to assessing productivity and quality, various breeding trait indices are historically used [8]. These parameters make it possible to group the existing varieties and lines into optimal clusters for further hybridization to obtain the model of the final variety conceived by the breeder [9]. Assessment for winter hardiness and frost resistance is one of the most important varietal traits for winter crops. The analysis of winter hardiness is one of the regular studies carried out by breeders [10]. Data collection is carried out with the onset of the time of the resumption of spring vegetation.

The assessment is carried out visually and is expressed in points from 0 to 5. This imposes a subjective factor on the final result and does not allow them to be included in statistical processing and reliably compare large amounts of experimental data. Only intuitively, the breeder chooses the best material, analyzing the points obtained in different years. In production, the crops, assessed by points 5 and 4, are left, and the heavily thinned ones (1…2 points) are subject to reseeding. The fate of crops, assessed by a score of 3, is more difficult to decide. If the preserved plants have grown well in autumn and can form 400–500 ears of corn per square meter, such sowing is preserved. If there are spots of dead plants in the crops (less than 50% of the field area), it is necessary to sow them on time with the same spring crop. With a uniform sparseness of production crops, a continuous overseeding of another crop (barley, spring wheat) on them, as a rule, does not give positive results. Such crops are either left, enhancing their care, or reseeded [11].

Remote sensing technologies have recently been used to solve problems using quantitative estimates of vegetation cover, for example, unmanned aerial vehicles (UAV) [12, 13]. In precision farming, drones can be used to identify areas of fields where it is necessary to quickly resolve current issues, for example, to carry out additional agricultural operations [14, 15]. This provides technicians with the ability to respond on time to any detected problem. [16]. Implementation of UAV solves many different tasks in precision farming [17]. Multispectral cameras are the most popular tool in agriculture [1, 18, 19]. Multispectral drone surveying allows you to calculate various vegetation indices, for example, Normalized difference vegetation index (NDVI) for assessing plant phytomass [20], Normalized difference Red Edge index (NDRE) – to assess the nitrogen content in plants [21], Chlorophyll Index Green (ClGreen) – to assess the content of chlorophyll. The data obtained from the UAV is more accurate and impartial than from hand-held scanners and has a higher resolution than from the satellite.

The study aimed at comparing the traditional methods of eye assessment of winter hardiness of breeding material in winter crops experiments with the possibility of remote sensing small areas with UAV, which makes it possible to increase the accuracy and objectivity of the breeder's experiment, eliminating the subjective factor in assessing the trait "overwintering" of breeding varieties and lines of winter cereals.

2 Materials and Methods

2.1 Study Area

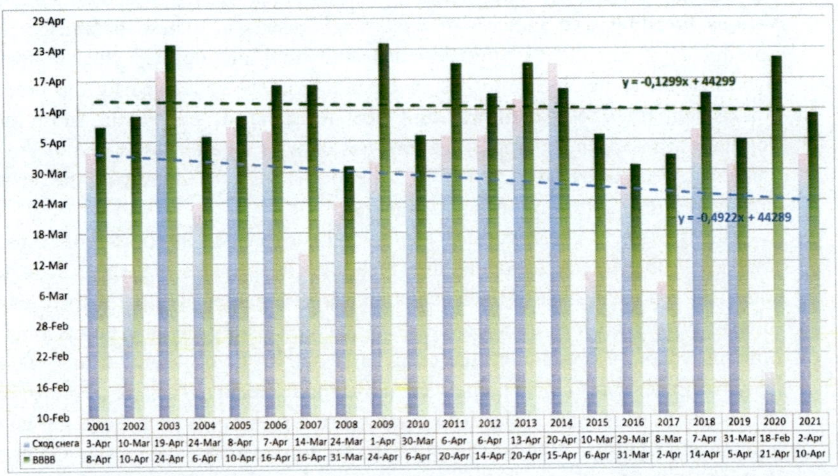

Fig. 1. Correspondence of the dates of snow melting and the onset of the time of the resumption of spring vegetation in the fields of FSBSI FSC

The study was carried out on the breeding field of winter crops of FSBSI FSC LGC, Oryol region, 53.013983, 35.990275. Analysis of the dates of snowmelt and the onset of the resumption of spring vegetation over the past 20 years has shown that with the trend of earlier snow melting, the date of the resumption of the spring vegetation of winter crops practically does not change and, on average, its onset is noted on April 12 (Fig. 1). Accordingly, the third decade of April was accepted as the optimal time for remote sensing of overwintered crops. The time of snowmelt and the resumption of the spring vegetation of winter crops in the region are described by the Eqs. (1) and (2), where y – number of days from January 1 of the forecast year, and x – the number of days until January 1 of the forecast year from 01.01.1900.

The entire breeding field was divided into 9 sections A-I (Fig. 2.). The total sowing area was 3.2352 ha.

Fig. 2. Location of breeding plots of winter crops

2.2 Ground Studies

Snowmelt in 2021 occurred on April 2. The time of the resumption of spring vegetation was noted on April 10, 2021. A visual score assessment of the state of overwintered crops was carried out on April 17. The methodology of state variety testing of agricultural winter crops provides for a five-point scale: 5 points - sparseness is imperceptible; 4 points - at least 70…80% of plants survived; 3 points - about 50% of the plants have survived; 2 points - less than 50% of plants survived; 1 point - 15…20% of plants have survived; the complete death of plants is estimated at 0. [11]. In the breeding process, reseeding or overseeding on plants that have fallen out over the winter is not used, but an independent indicator of the presence of the number and uniformity of overwintered breeding lines is necessary for analytical work.

2.3 Unmanned Aerial Vehicle Platform and Sensor

A DJI Matrice 200 v2 quadrocopter with an installed GNSS antenna was used during the research. The used quadrocopter proved to be a reliable research platform for various scientific purposes [22, 23]. A drone with dust and moisture protection is capable of flying at wind speeds up to 20 m/s, which is important when working in the field in open areas.

The quadrocopter is equipped with a modified DJI X4S 20Mp (5472 × 3648) camera with a three-axis stabilizer. The flights were carried out using the DJI Pilot mobile app. A MicaSense Altum multispectral camera with a DLS 2 sensor with a built-in GPS receiver was installed using a special suspension. The camera simultaneously takes pictures in six channels: blue (B) (475 nm center, 32 nm bandwidth), green (G) (560 nm center, 27 nm bandwidth), red (R) (668 nm center, 16 nm bandwidth), red edge (RE) (717 nm center, 12 nm bandwidth), near-IR (NIR) (842 nm center, 57 nm bandwidth) and LWIR (thermal infrared 8–14 μm). Resolution of spectral channels 3.2 MP (2064 × 1544) and thermal channel: 160 × 120.

EMLID Reach RS2 multi-frequency GNSS receiver was used to obtain high-precision data. The connection took place to the OREL base station in the Oryol region, located at a distance of less than 20 km (Fig. 3).

Fig. 3. Platform solution for high-precision monitoring of agricultural fields. In the picture from left to right: a remote control with a smartphone with installed mission planners, a UAV DJI Matrice 200 v2 UAV with an installed GNSS antenna and a MicaSense Altum multispectral camera with a light sensor, a multi-frequency GNSS receiver EMLID Reach RS2 on a tripod.

Ten control points (GCP) measuring 50 by 50 were used to check the accuracy of the data. The GCPs were evenly spaced across the field to accommodate elevation differences. The exact coordinates of the control points were recorded using a multi-frequency geodetic receiver EMLID Reach RS2, which was set in Survey mode.

2.4 Data Processing

Photogrammetric data processing was carried out using the Pix4DMapper software. To process RGB data, the 3D Maps template was used, while processing multispectral data, we used the Ag Multispectral template. As a result, a high-resolution orthophotomap was created, vegetation maps of three vegetation indices: Normalized difference vegetation index (NDVI), Normalized difference Red Edge index (NDRE), Chlorophyll Index Green (ClGreen). Indices are calculated using formulas (1)–(3), respectively:

$$NDVI = \frac{NIR - R}{NIR + R}, \quad (1)$$

$$NDRE = \frac{NIR - RE}{NIR + RE}, \quad (2)$$

$$ClGreen = \frac{NIR}{G} - 1, \quad (3)$$

3 Results

Monitoring from UAV was carried out at the end of the 3rd decade of April – 29.04.2021. The flight was at a height of 50 m, 429 photographs of the visible range and 816 photographs of each of the spectral ranges were collected. Longitudinal and lateral overlap for multispectral data was set to 75%. The resolution (GSD) for the orthophotomap was 1.37 cm/pix, for vegetation maps - 2.29 cm/pix.

As a result of the analysis of the obtained data, it was noted that three areas (*C, D, I*) are distinguished by lower average values of vegetation indices (Table 1). This is due to the fact that plots with clear vegetation zones between them are located in plots *C* and *D*, and the last plot *I* was laid at a late date (seedlings on November 7, 2020) and had not yet bush in the spring (Fig. 2. (*I*), Fig. 4a) in comparison to the laid down optimal sowing of the same winter wheat cultivar Valtorna (Fig. 2(H), Fig. 4b).

Table 1. Values of vegetation indices of the considered breeding plots

Plot	Culture	Area, ha	NDVI		NDRE		ClGreen	
			Mean	σ	Mean	σ	Mean	σ
A	Winter rye (PR-1)	0,08196924	0,80	0,17	0,39	0,09	4,15	1,39
B	Winter wheat (PR-1) breeding line B-20	0,47700768	0,78	0,14	0,42	0,09	4,60	1,56
C	Competitive test of 180 plots	0,26828908	0,52	0,31	0,28	0,16	3,08	2,39
D	Breeding Nursery 720 lines	0,07935739	0,40	0,25	0,22	0,10	1,96	1,34
E	Winter wheat (PR-1) breeding line No. 13	0,13911485	0,70	0,18	0,33	0,09	3,47	1,35
F	Winter wheat (PR-1) breeding line A-71	0,26668663	0,79	0,14	0,43	0,09	5,05	1,74
G	Winter wheat (PR-1) early sowing variety Valtorna	0,73487229	0,75	0,15	0,37	0,08	4,13	1,36
H	Winter wheat (PR-1) Valtorna variety for optimal sowing time	0,57241175	0,73	0,14	0,35	0,07	3,75	1,16
I	Winter wheat (PR-1) late sowing variety Valtorna	0,21343318	0,37	0,17	0,19	0,06	1,43	0,59

Fig. 4. The state of overwintered crops of winter wheat cultivar Valtorna at different sowing dates (a - October 20, 2020; b - September 23, 2020) at the time of the resumption of the growing season on April 12, 2021.

It should be noted that the standard deviations of the vegetation indices of sites with plots (Fig. 2 (*C*), Fig. 2 (*D*)) have an increased value (by 10…15%), what can be characterized by the presence of areas with no vegetation between the plots. The picture with late sowing (Fig. 2 (*I*)) is different - here, with low values of vegetation indices, the standard deviation corresponds to the uniformity of distribution of vegetative plants over the surface of the plot (the value is the same as in options for earlier sowing periods). This behavior of the standard deviation of the NDVI index can be used when assessing production crops, but requires a systematic approach, additional analysis and methodological experiments. The obtained data, selection of the optimal camera resolution and optimal differentiation by the area of minimal structural zones, training of neural networks will make it possible to quickly isolate problem areas with critical loss from industrial crops without the involvement of a mobile specialist agronomist (scout).

From the studied varieties and breeding lines (Fig. 2) on the site of competitive testing (Fig. 2 (*C*)), comparative experiments were methodically set up in 5-fold replication on the plots. Accordingly, in order to compare the visual (scoring) assessment of winter overwintering with the values of the vegetation index NDVI, the corresponding calculations (Fig. 5a) and visual assessment of the plots (Fig. 5b) in the breeding nursery were made (Fig. 2 (*C*)).

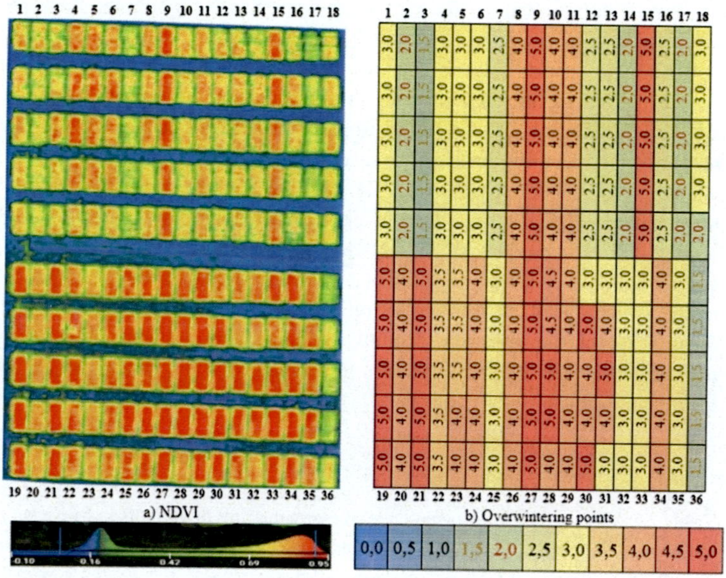

Fig. 5. NDVI vegetation index map (a) and overwintering score (b) of winter crops on plots in the FSBSI FSC LGC competitive trial, spring 2021 * Figures are numbers of options (1–36) for 5 repetitions

Correlation analysis on 180 plots of the correspondence of vegetation indices (NDVI, NDRE, ClGreen) and visual scores for overwintering revealed a significant relationship between them (Table 2), which characterizes the presence of a significant relationship between the compared data of vegetation indices and visually assessed overwintering of crops.

Table 2. Correlation coefficients between the vegetation indices and the visual assessment of winter crops overwintering at the time of the resumption of spring vegetation at the FSBSI FSC LGC breeding plot, spring 2021

	NDVI	NDRE	ClGreen	Points
NDVI	1			
NDRE	0,925955	1		
ClGreen	0,911714	0,985568	1	
Points	0,695909	0,721656	0,736366	1
	t_{05} = 1,97	<	t_{fakt} = 12,93	

Not all experiments in the competitive test had 5 repetitions, therefore, to identify differences in overwintering by analysis of variance, only methodically aligned 29 options were used (Table 3). The winter wheat variety Skipetr was taken as the reference cultivar in the experiment. This variety is a reference variety in the Central Black Earth Region and the Oryol Region, widespread in the Russian Federation and recommended as a winter-hardy variety [24]. Above standard (Fig. 5.a), highlighted 18 variants – 4, 5, 9, 11, 15, 19…22, 24…27, 29, 32…35. 4 variants significantly lost to the standard – 2, 7, 12, 16. At the level of standard were 6 variants – 3, 6, 8, 10, 13, 14. Variants 17, 18, 23, 28, 30, 31, 36 were composed of several lines and varieties (had one replication) and did not participate in the analysis (Table 3.). In general, according to the indicators of all the vegetation indices involved, calculated at the time of the resumption of the spring vegetation, it became possible to objectively assess the overwintering and identify varieties and breeding lines that significantly exceed the local standard.

Table 3. Average values of parameters of overwintering of variants in five repetitions (different varieties and breeding lines) at the moment of resumption of vegetation in the competitive nursery of winter crops FSBSI FSC LGC, 2021

№	Variant	Points	NDVI	NDRE	ClGreen
1	1. Skipetr (standard)	3	0,744	0,388	4,304
2	2. №13	2	*0,704*	*0,352*	*3,828*
3	3. №16	1,5	0,720	0,372	*3,896*
4	4. №17	3	**0,792**	**0,432**	**5,382**
5	5. №21	3	**0,778**	**0,420**	**4,784**
6	6. №22	3	0,760	**0,412**	4,592
7	7. L32	2,5	0,650	0,318	3,274
8	8. L982/08	4	0,776	0,396	4,386
9	9. A71	5	**0,820**	**0,466**	**5,950**
10	10. Timiryazevskaya Odnostebel'naya (Timiryazevskaya One-Stem)	4	0,736	*0,368*	*3,842*

(*continued*)

Table 3. (*continued*)

№	Variant	Points	NDVI	NDRE	ClGreen
11	11. Leonida x ABC	4	**0,788**	**0,408**	4,598
12	12. Aist x Fer17A	2,5	*0,698*	0,372	*3,800*
13	13. Aist x Fer17B	2,5	0,742	0,384	4,022
14	14. Sineva mut	2	0,726	0,372	*3,896*
15	15. Valtorna	5	**0,818**	**0,452**	**5,684**
16	16. M-39 x Sineva d.19	2,5	*0,688*	*0,362*	*3,662*
17	19. Leonida	5	**0,860**	**0,506**	**7,036**
18	20. Mezhdurechenka	4	**0,822**	**0,420**	**5,110**
19	21. Eritro K-8 x Fer-17	5	**0,864**	**0,508**	**6,884**
20	22. K/C 27/17	3,5	**0,836**	**0,458**	**5,796**
21	24. Tim. Yubileinaya. (52h-13)	4	**0,832**	**0,454**	**5,894**
22	25. Pochaevka x Leonida	3	**0,842**	**0,482**	**6,370**
23	26. M-39 x Sineva d.57	4	**0,878**	**0,474**	**6,160**
24	27. Skipetr x Leonida	5	**0,870**	**0,506**	**7,100**
25	29. Tritikale Alatyrskaya	4	**0,860**	**0,498**	**6,304**
26	32. Alexandra (51h-12)	3	**0,846**	**0,450**	**5,846**
27	33. 132h-15	3	**0,838**	**0,490**	**6,440**
28	34. 134h-15	4	**0,846**	**0,484**	**6,146**
29	35. 158h-16	3	**0,838**	**0,454**	**5,516**
By experience		3,45	0,7921	0,4296	5,189724
LSD_{05}*			0,03255	0,01961	0,39073

Italic - significantly below the standard (variety Skipetr), **Bold** - the indicators significantly exceeded the standard.

*LSD_{05} was assessed by the method of B.A. Dospekhov (field one-factor experiment) [25].

Subjectivity and laboriousness of traditional eye assessment of winter crops overwintering, large variability of weather factors in different years limit a scientist in the amount of the analyzed source material, because maintaining a wide variety of genetically heterogeneous varieties and lines on breeding crops is laborious in analysis, economically unprofitable, and makes it difficult to compare data from previous years. Transition to the analysis of the data of vegetation indices obtained at the time of the resumption of the vegetation, comparison to the reference variety in the region will allow the transition to an objective statistical assessment of the material and expand the volume of the analyzed breeding material on the basis of "overwintering" of winter crops of different years. Accumulation of an array of empirical (experimental) data in the form of structured digital information on the characteristic collection of winter cereals and compared to the data of the reference variety adapted in the region, in the future it can be used for inductive machine learning of neural networks [26]. Such specialized software will allow partially automating the breeding process, expanding the capabilities of the breeder and save funds.

4 Summary and Conclusions

1. The analysis of the dates of snow melting and the onset of the resumption of spring vegetation over the past 20 years showed that with the trend of earlier snow melting, the date of the resumption of the spring vegetation of winter crops practically does not change and on average the onset of the resumption of spring vegetation is noted on April 12 in the Oryol region. Accordingly, the third decade of April is recommended for collecting aerial photography data for assessing wintering of winter crops.
2. When assessing the overwintering of winter crops, a significant positive correlation ($r > 0.7$) was revealed between the vegetation indices (NDVI, NDRE, ClGreen) and the visual score in the initial period after the resumption of the spring vegetation. This allows us to objectively statistically compare the breeding material with each other based on "winter hardiness". The technology is recommended for methodical testing of varietal material of winter grain crops before transferring the variety for state testing.
3. It was noted that the standard deviation of the NDVI index ($r = 0.55$) more than other vegetation indices (NDRE, ClGreen) is associated with the degree of variegated crop loss (sparseness). This will make it possible to promptly signal problem areas when monitoring large areas, but additional research is required to optimize the resolution of cameras and the minimum elements of the treated areas.
4. The vegetation index data obtained in the initial period after the resumption of the spring vegetation of winter crops, tied to the regional standard reference variety, can be accumulated over the years and used for further machine learning of specialized neural networks.

References

1. Mulla, D.J.: Twenty five years of remote sensing in precision agriculture: key advances and remaining knowledge gaps. Biosys. Eng. **114**(4), 358–371 (2013)
2. Arakeri, M.P., Arun, M., Padmini, R.K.: Analysis of late blight disease in tomato leaf using image processing techniques. Int. J. Eng. Manuf. (IJEM) **5**(4), 12–22 (2015). https://doi.org/10.5815/ijem.2015.04.02
3. Nair, T., Singh, A., Venkateswarlu, E., Swamy, G.P., Bothale, V.M., Krishna, B.G.: Generation of analysis ready data for Indian Resourcesat sensors and its implementation in cloud platform. Int. J. Image Graph. Sig. Process. (IJIGSP) **11**(6), 9–17 (2019). https://doi.org/10.5815/ijigsp.2019.06.02
4. Govaerts, B., Verhulst, N.: The normalized difference vegetation index (NDVI). GreenSeeker handheld sensor: toward the integrated evaluation of crop management. Part A – Concepts and case studies, p. 12 (2010)
5. Zhelezova, S.V., Ananiev, A.A., Vyunov, M.V., Berezovsky, E.V.: Monitoring of winter wheat crops using unmanned aerial photography and an optical sensor GreenSeeker® RT200. Bull. Orenburg State Univ. **6**(194), 56–61 (2016)

6. Lobachevsky, Ya.P., Dorokhov, A.S.: Promising scientific and technical projects in the field of mechanization and robotization of agriculture. In: Formation of a Single Scientific and Technological Space of the Union State: Problems, Prospects, Innovations, pp. 333–343 (2017)
7. Lachuga, Yu.F., Izmailov, A.Yu., Lobachevsky, Ya.P., Shogenov, Yu.Kh.: Development of intensive machine technologies, robotic equipment, efficient energy supply and digital systems in the agro-industrial complex. Mach. Equip. Village **6**(264), 2–9 (2019)
8. Boroevich, S.: Principles and methods of plant breeding, p. 344 (1984)
9. Tugareva, F.V., Sidorenko, V.S., Vilyunov, S.D., Malchikov, P.N., Myasnikova, M.G.: The use of cluster analysis in identifying valuable breeding material of interspecific hybrids of spring wheat (Triticum durum x Triticum dicoccum). In: Materials of the International Scientific and Practical Conference of Young Scientists and Specialists, pp. 159–161 (2019)
10. Zadorin, A.M., Zotikov, V.I., Zelenov, A.A., Sidorenko, V.S., Budarina, G.A., et al.: Recommendations for conducting spring field work in the Orel region in 2020, p. 56 (2020)
11. Golovachev, V.I., Kirillovskaya, E.V.: Methods of state variety testing of agricultural crops. Issue two. Cereals, cereals, legumes, corn and forage crops, p. 194 (1989)
12. Cheng, X., Wang, J., Xu, Y.: A method for building a Mosaic with UAV images. Int. J. Inf. Eng. Electron. Bus. (IJIEEB) **2**(1), 9–15 (2010)
13. Kurbanov, R.K., Zakharova, N.I.: Application of vegetation indexes to assess the condition of crops. Agric. Mach. Technol. **14**(4), 4–11 (2020)
14. Ramesh, K.N., Chandrika, N., Omkar, S.N., Meenavathi, M.B., Rekha, V.: Detection of rows in agricultural crop images acquired by remote sensing from a UAV. Int. J. Image Graph. Sig. Process. (IJIGSP) **8**(11), 25–31 (2016). https://doi.org/10.5815/ijigsp.2016.11.04
15. De Camargo, T., Schirrmann, M., Landwehr, N., Dammer, K.-H., Pflanz, M.: Optimized deep learning model as a basis for fast UAV mapping of weed species in winter wheat crops. Remote Sens. **13**(1704) (2021). https://doi.org/10.3390/rs13091704
16. Kurbanov, R.K., Zakharova, N.I., Gaiduk, O.M.: Using the heat channel (LWIR) to assess the state of crops and forecast the yield of agricultural crops. Electrotechnol. Electr. Equip. Agro-Ind. Complex **67**(3), 87–94 (2020)
17. Lichman, G.I., Lobachevsky, Ya.P., Elizarov, V.P., Kurbanov, R.K.: Use of UAVs for monitoring the state of breeding sites. In: Scientific and Information Support of Innovative Development of the Agro-Industrial Complex, pp. 311–315 (2017)
18. Cui, B., Zhao, Q., Huang, W., Song, X., Ye, H., Zhou, X.A.: New integrated vegetation index for the estimation of winter wheat leaf chlorophyll content. Remote Sens. **11**(974) (2019). https://doi.org/10.3390/rs11080974
19. Tsouros, D.C., Bibi, S., Sarigiannidis, P.G.: A review on UAV-based applications for precision agriculture. Information **10**(11) (2019). https://doi.org/10.3390/info10110349
20. Yue, J., Feng, H., Li, Z., Zhou, C., Xu, K.: Mapping winter-wheat biomass and grain yield based on a crop model and UAV remote sensing. Int. J. Remote Sens. **42**(5), 1577–1601 (2021)
21. Zillmann, E., Schönert, M., Lilienthal, H., Siegmann, B., Jarmer, T., Rosso, P., Weichelt, H.: Crop ground cover fraction and canopy chlorophyll content mapping using RapidEye imagery. Int. Arch. Photogram. Remote Sens. Spat. Inf. Sci. **XL-7/W3**, 149–155 (2015). https://doi.org/10.5194/isprsarchives-XL-7-W3-149-2015
22. Daugela, I., Visockiene, J.S., Kumpiene, J.: Detection and analysis of methane emissions from a landfill using unmanned aerial drone systems and semiconductor sensors. Detritus **10**, 127–138 (2020). https://doi.org/10.31025/2611-4135/2020.13942

23. Beranek, C.T., Roff, A., Denholm, B., Howell, L.G., Witt, R.R.: Trialling a real-time drone detection and validation protocol for the koala (Phascolarctos inereus). Aust. Mammal. **43** (2), 260–264 (2020). https://doi.org/10.1071/AM20043
24. Soft winter wheat Scepter variety (2009). https://reestr.gossortrf.ru/sorts/9553093/. Accessed 14 June 2021
25. Dospekhov, B.A.: Methodology of field experience (with the basics of statistical processing of research results), p. 351 (1985)
26. Makarenko, A.V.: Deep neural networks: origin, formation, current state. Manag. Probl. **2**, 3–19 (2020). https://doi.org/10.25728/pu.2020.2.1

The Hybrid Detection Methodology of Attacks for 5G

Maksim Iavich(✉)

Caucasus University, P. Saakadze st.1, 0102 Tbilisi, Georgia
miavich@cu.edu.ge

Abstract. The telecommunications industry experiences a fundamental transition towards 5G networks. Based on our research, it can be concluded that 5G still faces security challenges. In the frame of our analysis, we have determined several security concerns. It must also be mentioned that lately, the scientists have found different security problems in the existing 5G networks that give hacker the opportunity to install the undesirable code into the 5G system and perform undesirable actions. The corresponding attacks are MNmap, MiTM and the Battery drain attack. Therefore, it is mandatory to design new architectures for 5G and the future networks, which can be defined as 6G ones, for this, it is obligatory to provide the new machine learning algorithms, which will give the scientists the opportunity to provide high-level security. The article analyzes the existing security problems of 5G architecture. Based on the research, a new cyber security model is offered, which is based AI/ML algorithms. It envisions the fundamental components, the firewall, intrusion detection and intrusion protection systems. We apply the offered security model into already existing architecture of 5G. The paper essentially describes an Intrusion Detection System (IDS), which is a part of proposed model. The IDS implements a comprehensive methodology for the identification of the attacks to which the 5G systems can be vulnerable. These attack patterns include MNmap, MiTM and the Battery drain attacks. The methodology is described in the paper. The article analyzes the efficiency of the described model, through the design and implementation of a comprehensive testing laboratory. In the laboratory there are a server and 50 raspberry pi, with fifty 4G modems, which are needed in order to make the simulation of the attacks on the corresponding server. The paper also suggests the improvement approach, which could be integrated into the system in future.

Keywords: 5G security · Attacks · Detection methods · Attacks on 5G

1 Introduction

The number of devices connected to the wireless network is growing unprecedentedly, and the volume of transmitted data is directly proportional to the number of devices. The introduction of 5G networks has been put on the agenda to meet the already existing and forecasted requirements. Development of new infrastructures determines the occurrence of new security hazards. Thus, the attacks like denial of service (DoS), distributed denial of service (DDoS), spoofing, man in the middle, and application layer attacks represent traditional problems of the global Internet network, and they can

Z. Hu et al. (Eds.): AIPE 2021, LNDECT 119, pp. 65–74, 2022.
https://doi.org/10.1007/978-3-030-97064-2_7

manifest themselves on next generation infrastructures, such as the 5G networks. The paper is structured according to the following sections. The next section discusses on the relevant technical concepts, and it also provides a brief overview of 5G network research. Moreover, the security problems are discussed, and the technical model of the proactive intrusion detection core is described, together with the types of the detected attack patterns. Furthermore, the training and the detection methodologies are described. Additionally, the system's real world performance and detection accuracy are assessed considering a dedicated testing infrastructure. The conclusion is the last chapter of the article.

2 Conceptions and Existing Materials

Today's technologies have made it possible to refine the architecture and layout methodology of the IoT [1–5]. Vodafone's cellular IoT [6] and 3GPP-based networking standards are also set. 5G networks will be able to provide IoT networks of billions of smart devices with secure transmission channels. In presented paper also are considered few WSDN (wireless software defined networking) models for 5G, such as SoftAir, CloudRAN and CONTENT [9]. As they show appropriateness of 5G. End users are supplied with service over the IoT network. The correct development of appropriate standards and technologies presupposes security and confidentiality aspects that pose new challenges for standards organizations as well as for government agencies [10, 11]. New 5G standards can increase the importance and spread of IoT networks, making it readily available on the basis of modern technology [12–14]. Moreover, 5G data networks must ensure the joint operation of IoT devices using different intelligent sensors [15–18]. The introduction of 5G networks should lead to an increase in the scale of IoT networks, as it is precisely based on these networks that provide low latency, reliability and optimal capacities. In the last few years, 5G IoT networks have been extensively explored in both academic and industrial settings [19]. Thus, the full introduction of 5G networks is expected once the relevant standards are fully specified by 2025. In [20] describes a heterogeneous computing platform for applications running on 5G IoT networks. Article [21] describes energy management solutions in the context of spectral solutions. In addition, studies are described connected to solving narrowband problems. The article [22] describes NB-SCMA (new narrowband sparse code multiple access) solution for 5G IoT uplinks. A sample radio channel for 5G IoT networks is described, which meets the requirements and its purpose is to provide efficient wireless communication. Also a model for energy efficient 5G IoT networks is described. The part [24] describes the energy efficiency and comprehensive architecture of these networks, including wireless and wired components. It also describes studies conducted by academic and industry groups on the aspect of 5G and IoT networks. Their purpose is to present the latest research and evaluations, the theoretical basis, the design of the attachments, and the standardization and implementation of all 5G IoTs. In addition, leading organizations such as Cisco, Intel and Verizon have jointly created a research project aimed at defining the 5G standard. The ultimate goal is to develop algorithms on the basis of neuroscience, which will be able to adapt the transmitted video materials to the capabilities of the human eye. Based on this, certain capabilities of 5G networks may help to enhance human intelligence. 5G technologies will be able to support the progress of IoT networks that will connect billions of smart devices and be able to link and share data without human intervention. Numerous existing IoT systems today provide quality of life improvements through communication between smart devices that operate in their everyday environment. Industrial IoT (IIoT) is in development and facing the need to address a variety of issues. This includes requirements for the transformation of business models and the design of the products involved. Systems with automated traffic engineering still face numerous technical issues related to safety and latency. New 5G technologies must also ensure the security of IoT networks and work without interruptions and latency.

2.1 Virtualized Wireless Network Function

The VWNF (virtualized wireless network function) is an additional pattern in 5G network frames. Which consequentially defines an autonomous 5G network over NFV (Network Function Virtualization) [10] on appropriate infrastructure of hardware. So, it allows the deployment of 5G IoT networks in properly configured cloud infrastructures [11]. It enables the deployment of 5G IoT networks on a cloud infrastructure that is correctly configured with the help of a virtual environment, which provides scale and versatility. It also ensures the creation of data networks that share network resources in a cloud environment. NFV ensures creation of independent networks tailored to the requirements of specific applications. This model can provide 5G IoT networks with real-time processing capability by optimizing data processing and transmission capacity and proper data transfer speeds. A network-like model is described in [13]. Also, several studies have confirmed that it is practically possible to create 5G networks specially optimized for the applications deployed in them. In addition, virtualization can optimize the use of radio resources used by IoT devices to connect to the data network. This is described in section [7], which also discusses current network virtualization solutions. Thus, the 5G IoT infrastructure will be able to support the required number of devices and related software applications. The cyber security model described in this document views the virtualization of 5G data networks as a fundamental design mechanism that allows appropriate security policies to be properly designed and implemented.

3 Security Problems of 5G and Description of Approach

The existing research contributions show that the 5G ecosystem is affected by security problems [25–27]. Thus, this paper discusses on the following essential aspects: 1) The 5G network is quite vulnerable to software attacks and it has much more penetration points as it mainly depends on the software configurations of its own logic architecture. Attackers can take advantage of bugs and vulnerabilities in the security system. 2) Because the 5G systems have new functionality, some network equipment components and functions are not yet properly protected. Objects of attacks can become both base stations as well as basic network management functions; 3) Mobile network providers are dependent on suppliers, which can also inspire new directions of attack. 4) Most of the extremely important IT applications will be involved in the 5G network, and in this case, these applications themselves can become the target of hackers. 5) 5G network users will also have plenty of devices, which may increase the ability to carry out various types of DoS and DDoS attacks; 6) Security issues can also be caused by network segmentation. There is a number of vulnerabilities in 5G security systems that can be exploited to inject malicious code and perform illegal activities. The hardware architecture of the proposed system involves the cybersecurity module to be integrated into 5G stations as an additional server, which includes a cyber security module that integrates a firewall and IDS/IPS (intrusion detection system/intrusion prevention system). The general architecture of the system is illustrated in Fig. 1.

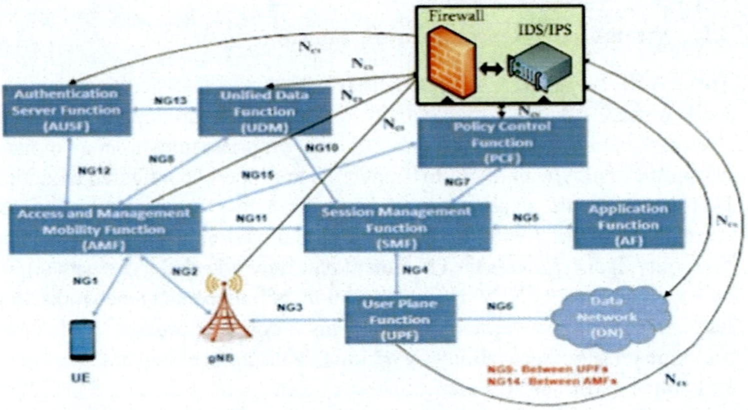

Fig. 1. The module of cyber security

Based on the research described in this article, a number of weaknesses of 5G networks are revealed. It is clear that it is weakly protected from Dos, Probe and software attacks. In the primary version of IDS presented by us, machine learning algorithms are applied. Different sets of data are used in the learning process to protect against these types of attacks. One of them is NSL KDD [28–31]. It is noteworthy that this set is often used in academic researches and to create prototypes of IDSs. Thus, this set of data is very often used to measure the detection of anomalies. In the next sections we offer the methodology of the mitigation techniques in order to protect against MNmap, MiTM and Battery drain attacks. The methodology is integrated into the novel IDS.

The MitM Attack

It was found out, that a hacker can perform a MitM (Man in the middle) attack on the 5G infrastructure by spoofing a base station, which causes users to connect to the fake station. Thus, the hacker`s intermediate station can forward the traffic to the legitimate base station or to the mobile network's core layer, which results in a classic MitM scenario, just like if we had spoofed a Wi-Fi access point`s ESSID and BSSID forcing the unsuspecting users to connect to the fake access point. This can result into the actual sniffing and reading/parsing of all the traffic going through the fake base station. The disclosure of private information is not the worst consequence, as the hacker may also try to change the transmitted information on the fly. One additional problem could be represented by the hackers trying to pivot from the base station to the mobile network's core segment, or at least to probe it for unknown vulnerabilities. MiTM is also the reason of MNmap, and the Battery drain attacks. The process is illustrated in Fig. 2.

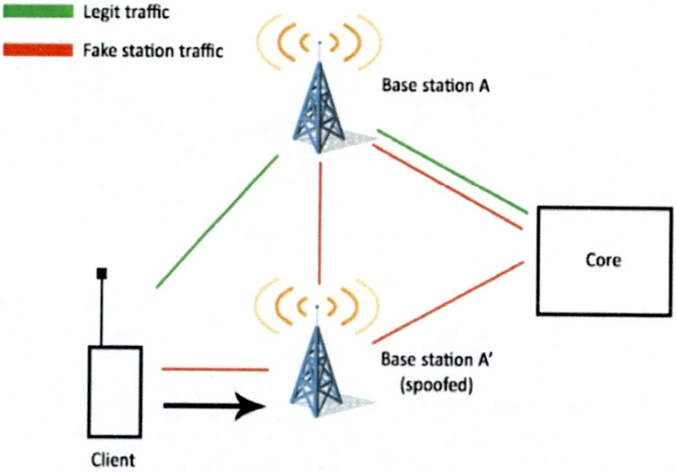

Fig. 2. The MitM attack

The Mitigation Solution

The most logical solution to mitigate MitM problem would be to prevent hackers from creating a fake base station, which can be used to spoofing the identity of a legitimate base station. Both of them are being done by manipulating information at the second layer of the OSI model. For example, let us consider two Wi-Fi access points, a legitimate one and the spoofed one. Their network information may look like this. **Legit AP:** BSSID - AA:BB:CC:DD:EE:FF; ESSID – TEST; **Spoofed AP:** BSSID - AA:BB:CC:DD:EE:FF; ESSID – TEST. It can be noticed that the network addresses are the same. Therefore, the goal is to identify whether the transmitted access point's identity is spoofed. The possible theoretical solution would be to consider one more antenna, which will constantly work in a monitoring mode, and therefore send information about neighboring access points to the relevant software, which would parse the acquired information and seek for coincidences between the real ESSID/BSSID, and the information obtained from the scanner. Thus, if it finds the same ESSID or BSSID, this implies that the legitimate access point's identity was spoofed. Nevertheless, this approach is not suitable in the case when the spoofed device is out of the antenna`s range, as it will be physically difficult to detect, although this may be fixed by using a rotating directed antenna or a higher power isotropic antenna. The same approach can be used in order to detect spoofed 5G base stations. Furthermore, the fake base stations can be detected by creating a list of legitimate entities, which is shared across all base stations. Thus, if a base station broadcasts itself, which is not part of the whitelist, then it may be safely inferred that it spoofs the identity of a legitimate base station. The process is shown in Fig. 3.

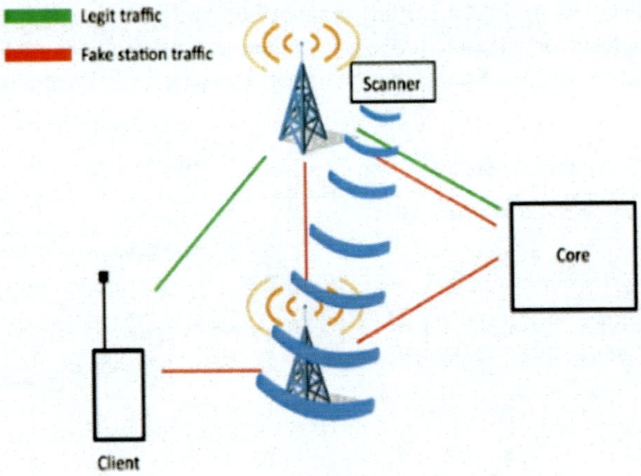

Fig. 3. Mitigation model

4 Methodology

During the training, two DOS datasets were used, and we have implemented the mitigation methodology against MNmap, MiTM and battery drain attacks. The NSL-KDD data set is a rather well known benchmark used in the research of the different Intrusion Detection techniques. The proper training of the dataset considers the selection of the classification algorithm, and the creation of the multilayer neural network. We have divided the NSL-KDD data set into two parts. 90% of the information is collected in the first - training data set, and the remaining 10% in second- test data set. The DOS and DDOS data sets are also divided into two parts, respectively. Where 80% of the data is collected in the study part and, consequently, 20% in the test part. The method of dividing the data in this way resulted in high accuracy. The model of the training was conducted in different stages, using one database at a time. The precision of the model prepared with NSL-KDD data set is 0.9711049372916336, while with DOS it is 0.9998861703182078. Upon completion of the training process, the system expects input data from the network component. Initially, the input is checked for attacks in NSL-KDD database. Once an attack pattern is identified, it is passed to an IPS. In the case of samples in the DOS database, it is checked in a parallel stream. In the event of an attack pattern being detected, it passes it to the IPS component. The occurrence of the MNmap, MiTM and Battery drain attacks is checked in the thread in the parallel mode, it is done by checking the MiTM attack. In the case of detecting attack pattern, the information is passed over to the IPS components for further processing. In addition, given that no pattern of attack a has been detected, the IDS system detects that the traffic is clean and starts to work on the next unit.

5 Experiments

The intrusion detection system was tested considering a test laboratory using 50 RASPBERRY PI devices. Consequently, we have used the same number of modems with 4G SIM cards. The intrusion detection system was installed on the server. In the study of generated traffic, we focused on attacks that were used by means of network sniffer; in the process we analyzed the corresponding parameters of all NSL-KDD and DOS patterns using the Python programming language. We converted the output to the original data set format. We sent all this information into the IDS system. Obtained results are presented in the Table 1.

Table 1. IDS precision

Attack type	Number of attacks	Identified attacks
LAND	100	100
BACK	100	99
POD	100	100
NEPTUNE	100	98
SMURF	100	96
TEARDROP	100	82
BUFFER_OVERFLOW	100	76
FTP_WRITE	100	86
ROOTKIT	100	62
LOADMODULE	100	91
GUESS_PASSWD	100	100
MULTIHOP	100	91
SPY	100	51
PROBE	100	98
IPSWEEP	100	92
NMAP	100	95
PORTSWEEP	100	98
SATAN	100	82
MSSQL	100	99
LDAP	100	98
NetBIOS	100	98
Syn	100	100
Portmap	100	97
Man in the middle	20	20

The results prove that the IDS is rather accurate, and it can be used as the prototype version of the future real world system. It can be noticed that the system can identify DOS/DDOS attacks with high levels of accuracy, as a consequence of the additional DOS/DDOS dataset that was considered. The system also identified the man in the middle attack patterns with the highest possible accuracy. Therefore, the table shows that our approach identifies DOS/DDOS better than the existing ones and it can also protect against man in the middle attacks with high accuracy.

6 Experimental Results and Discussion

The research that is reported in this paper created the novel intrusion detection system, which is specifically tailored for 5G networks attacks. The IDS considers machine learning algorithms, which are trained using the NSL-KDD dataset and the DOS/DDOS attacks dataset. Furthermore, it is relevant to assert that the IDS is trained using the relevant attack vectors, which represent major vulnerabilities for 5G networks. We have also analyzed the MNmap, MiTM and battery drain attacks and explained the mechanisms through which these attacks are successful. Consequently, proper protection models were designed and implemented into the intrusion detection system.

As a consequence, the system is able to detect the relevant 5G networks attack patterns in a real time fashion, while the system's performance is assessed using the described test infrastructure. It must be mentioned that it is not efficient to receive data in NSL-KDD format, so in future it would be interesting to use the own attack patterns for training.

7 Conclusion

The performance assessment process determined that the novel intrusion detection system efficiently detects the attack patterns, better then classical not hybrid approaches. Furthermore, the system's efficiency can be improved through the implementation of parallel processing routines in its main processing core. The development team is actively working on the proper design and implementation of this necessary reengineering. Additionally, the machine learning component will be trained using the relevant data of the attack patterns. It is relevant to note that the intrusion detection system can also protect the secured 5G mobile network against the most relevant attack patterns, such as MNmap, MiTM and battery drain attacks. The relevant proactive detection components will be improved in order to offer optimal security mechanisms for the intended real world 5G network structures.

References

1. Liu, Y., Qin, Z., Elkashlan, M., Ding, Z., Nallanathan, A., Hanzo, L.: Non-orthogonal multiple access for 5G and beyond. Proc. IEEE (2018)
2. I-Scoop, 5G and IoT in 2018 and beyond: the mobile broadband future of IoT. https://www.i-scoop.eu/internet-of-things-guide/5g-iot/

3. Akpakwu, G.A., Silva, B.J., Hancke, G.P., Abu-Mahfouz, A.M.: A survey on 5G networks for the internet of things: communication technologies and challenges. IEEE Access **6**, 3619–3647 (2017)
4. Nunez, M.: What Is 5G and How Will It Make My Life Better? https://gizmodo.com/what-is-5g-and-how-will-it-make-my-life-better-1760847799
5. The Tech Wire Asia: The next generation of IoT. http://techwireasia.com/2017/08/next-generation-iot/
6. Parvez, I., Rahmati, A., Guvenc, I., Sarvat, A.I., Dai, H.: A survey on low latency towards 5G: RAN, core network and caching solutions. IEEE Commun. Surv. Tutor. **20**(4), 3098–3130 (2018)
7. Xia, X., Xu, K., Wang, Y., Xu, Y.: A 5G-enabling technology: benefits, feasibility, and limitations of in-band full-duplex mMIMO. IEEE Veh. Technol. Mag. **13**(3), 81–90 (2018)
8. Wu, J., Zhang, Z., Hong, Y., Wen, Y.: Cloud radio access network (C-RAN): a primer. IEEE Netw. **29**(1), 35–41 (2015)
9. Project CONTENT FP, 2012–2015. http://cordis.europa.eu/fp7/ict/future-networks/
10. SDX Central: How 5G NFV Will Enable the 5G Future, 15 January 2018. https://www.sdxcentral.com/5g/definitions/5g-nfv/
11. Kaplan, K.: Will 5G wireless networks make every internet thing faster and smarter? https://qz.com/179794/will-5g-wireless-networks-make-every-internet-thing-faster-and-smarter/
12. Hosek, J.: Enabling technologies and user perception with integrated 5G-IoT Ecosystem (2016)
13. Chen, M., Qian, Y., Hao, Y., Li, Y., Song, J.: Data-driven computing and caching in 5G networks: architecture and delay analysis. IEEE Wirel. Commun. **25**(1), 70–75 (2018)
14. Stephan, J., Krishnamurthy, K.: Understanding the industrial internet of things. http://usblogs.pwc.com/emerging-technology/understanding-the-industrial-internet-of-things/
15. BLE: Smart Bluetooth Low Energy. http://www.bluetooth.com/Pages/Bluetooth-Smart.aspx
16. Nokia: LTE Evolution for IoT Connectivity. Nokia, Technical Report, Nokia White Paper, pp. 1–18 (2016)
17. RPMA: RPMA Technology for the Internet of Things. Ingenu, Technical Report (2016)
18. SigFox: SigFox. http://www.sigfox.com
19. Costanzo, A., Masotti, D.: Energizing 5G. IEEE Microw. Mag. (2017)
20. Schinianakis, D.: Alternative security options in the 5G and IoT era. IEEE Circuits Syst. Mag. **17**(4), 6–28 (2017)
21. Boulogeorgos, A.-A.A., et al.: Terahertz technologies to deliver optical network quality of experience in wireless systems beyond 5G. IEEE Commun. Mag. **56**(6), 144–151 (2018)
22. Khal, B., Hamdaoui, B., Guizani, M.: Extracting and exploiting inherent sparsity for efficient IoT support in 5G: challenges and potential solutions. IEEE Wirel. Commun. **24**(5), 68–73 (2017)
23. Xu, L., Collier, R., O'Hare, G.M.P.: A survey of clustering techniques in WSNs and consideration of the challenges of applying such to 5G IoT scenarios. IEEE Internet Things J. **4**(5), 1229–1249 (2017)
24. Sekander, S., Tabassum, H., Hossain, E.: Multi-tier drone architecture for 5G/B5G cellular networks: challenges, trends, and prospects. IEEE Commun. Mag. **56**(3), 96–103 (2018)
25. Iavich, M., Gnatyuk, S., Odarchenko, R., Bocu, R., Simonov, S.: The novel system of attacks detection in 5G. In: Barolli, L., Woungang, I., Enokido, T. (eds.) AINA 2021. LNNS, vol. 226, pp. 580–591. Springer, Cham (2021). https://doi.org/10.1007/978-3-030-75075-6_47
26. Bocu, R., Iavich, M., Tabirca, S.: A real-time intrusion detection system for software defined 5G networks. In: Barolli, L., Woungang, I., Enokido, T. (eds.) AINA 2021. LNNS, vol. 227, pp. 436–446. Springer, Cham (2021). https://doi.org/10.1007/978-3-030-75078-7_44

27. Lv, Z., Singh, A.K., Li, J.: Deep learning for security problems in 5G heterogeneous networks. IEEE Netw. **35**(2), 67–73 (2021). https://doi.org/10.1109/MNET.011.2000229
28. Sakr, M.M., Tawfeeq, M.A., El-Sisi, A.B.: An Efficiency optimization for network intrusion detection system. Int. J. Comput. Netw. Inf. Secur. (IJCNIS) **11**(10), 1–11 (2019). https://doi.org/10.5815/ijcnis.2019.10.01
29. Manjunatha, B.A., Gogoi, P., Akkalappa, M.T.: Data mining based framework for effective intrusion detection using hybrid feature selection approach. Int. J. Comput. Netw. Inf. Secur. (IJCNIS) **11**(8), 1–12 (2019). https://doi.org/10.5815/ijcnis.2019.08.01
30. Iqbal, A., Aftab, S.: A feed-forward and pattern recognition ANN model for network intrusion detection. Int. J. Comput. Netw. Inf. Secur. (IJCNIS) **11**(4), 19–25 (2019). https://doi.org/10.5815/ijcnis.2019.04.03
31. Tavallaee, M., Bagheri, E., Lu, W., Ghorbani, A.A.: A detailed analysis of the KDD CUP 99 data set. In: 2009 IEEE Symposium on Computational Intelligence for Security and Defense Applications, pp. 1–6 (2009). https://doi.org/10.1109/CISDA.2009.5356528

Islanding and Grid Disturbance Detection Based on Multi-SVMs Delay Validation in Synchronous Dominated Microgrids

Dan Zhou[1], Meiling Deng[2], Xu Deng[1], Wenfeng Wang[1], Jiajie Li[1], and Likun Chen[3(✉)]

[1] Shaoguan of Guangdong Power Grid Co., Ltd, ShaoGuan 512031, China
[2] Shaoguan Renhua of Guangdong Power Grid Co., Ltd, ShaoGuan 512399, China
[3] School of Electrical Engineering, Wuhan University, Wuhan 430072, China
chenlk@whu.edu.cn

Abstract. Intrinsically active, passive, or a hybrid method of the two is the main methods for solving the problem of islanding detection in microgrids. For islanding and disturbance detection, most methods rely on the common coupling point of microgrid and large power grid. Based on multiple support vector machines, this paper proposes an islanding and interference detection method. First, analyze the on-site electrical quantities and form feature vectors before training the support vector machine on the islanding and interference conditions; second, train the support vector machine on the islanding and interference conditions; finally, through the complementarity of multiple classifiers, improve islanding detection accuracy. The simulation example considers interference in multiple scenarios to validate the effectiveness of the above method, and the results show that the in-situ detection method in this paper gains robustness.

Keywords: Microgrid · Support vector machine · Islanding detection · Data mining · Feature vector

1 Introduction

With the increasingly prominent energy problem, microgrid becomes the inevitable development trend of power system. In the process of microgrid operation, island operation may cause unpredictable damage to human life and power grid, so island detection is one of the important problems that must be solved in the operation of microgrid.

Islanding detection is usually divided into active, passive and mixed methods. Reference [1] innovatively proposed two-level Islanding detection method based on active interference method. Set thresholds for the Rate of Change of Voltage (ROCOV) and Rate of Change of Active Output (ROCOP) for point of Common Coupling (PCC). This method reduces the non-Detection zone (NDZ) and improves the accuracy of island detection. However, like other active methods, this method will deteriorate the power quality and cost higher than passive detection methods.

Z. Hu et al. (Eds.): AIPE 2021, LNDECT 119, pp. 75–84, 2022.
https://doi.org/10.1007/978-3-030-97064-2_8

Comparatively speaking, passive detection does not cause disturbance to the system, but due to the similarity of isolated and non-isolated time domain signals, the detection accuracy is difficult to improve. Therefore, some references have proposed a method based on data-driven, i.e. Support Vector Machine (SVM) for passive island detection. Reference [2, 3] uses multiple SVMs to directly identify islands and faults, and achieves better detection accuracy through direct complementarities of multiple SVMs. Reference [4] evaluated the performance of SVM method with multiple feature inputs, providing important reference value for SVM-based islanding detection. In reference [5–9], feature extraction is performed on voltage and current data, and multiple feature vectors are obtained and placed into SVM to complete island detection. This is because there is little contrast between voltage and current changes in isolated and non-isolated conditions, so signals need to be decomposed. This method is efficient and accurate. However, these papers do not compare other electrical quantities such as frequency, rotor Angle, active power and reactive power, so higher accuracy detection may be missed. Non-detection zone is seen as an important concept in islanding detection and represents the mismatch in active and reactive power generated by the synchronous machines and consumed by the load and a lot of research is trying to minimized the NDZ with a higher accuracy [10–15].

The key to improve the detection accuracy is how to efficiently use the electrical signals of isolated islands and non-isolated islands to extract effective feature vectors under different conditions. Therefore, a SVM-based method was proposed in this paper. The voltage and three-phase root mean square, current and three-phase root mean square, frequency, active power and reactive power of distributed generation (DG) were selected as feature vectors to be put into SVM training.

Other than inverter-based generation, hydropower is power derived from the energy of falling water and running water and mostly based on synchronous generation. Synchronous generators meet with relatively higher capacity of speed regulation and frequency modulation and operate with less harmonic contents compared to the inverter-based generation, which makes it tough to discriminate the operating state because of the similiar voltage and frequency waveform between islanding and non-islanding. Generally, the carriers for islanding detection algorithm implementation are the primary stations or diverse distribution terminal units using the electrical parameters sampled by potential transformer (PT) and current transformer (CT). The essential components used for computing the sampled points are general purpose chips for the power industry. Consequently, power-specific integrated circuit (PSIC), which is evolved from application-specific integrated circuit (ASIC) and dedicated to solve the conundrum in power grid, gains momentum and exemplary application of PSIC was carried out in China. In the presence of an islanding situation, the PSICs, as the core computing infrastructure of the inspection equipment, are expected to perform their functions efficiently and within the prescribed time period to guarantee the safety of people and grid equipment, as well as power quality.

The results were compared and analyzed, and multiple SVMs were complementary to form the final detection result. The classification accuracy of this result can accurately detect the islanding.

2 Support Vector Machine-Based Islanding Detection Method

2.1 Basic Principles and Mathematical Derivation of Support Vector Machine

As a classic in the machine learning algorithm for solving classification problem with good effect, the current in the electric power industry in load forecasting, fault diagnosis, and other fields use is more, for the islanding detection, the core of the SVM is based on the data driven way of machine learning, in order to construct a hyperplane method separated island and the islanding of the eigenvector of cases, This is the basis for islanding detection.

For the classification problem, the primal problem can be present as formula (1):

$$\begin{cases} \max\limits_{\omega,b} \min\limits_{x_i} \frac{1}{\|\omega\|} |\omega^T x + b| \\ s.t\ \ y_i(\omega^T x + b) \geq 0 \end{cases} \tag{1}$$

Where, $\mathbf{x}_i \in \mathbf{R}^d$, $y_i \in \{-1, 1\}$ is the label classification, the core idea of SVM is biggest, make space equivalent to make $\|\omega\|$ to find the smallest classification interval of the optimal classification plane.

At this point, the original problem can be transformed into a dual problem, and the dual problem can be optimized by introducing Lagrange function. For island detection, the characteristic vectors of electrical gas input in the process of island isolation are usually nonlinear with each other, which leads to the need of kernel function of SVM for further feature vector processing, so the optimized duality problem will be transformed into (2):

$$\phi(\mathbf{w}, \xi) = \frac{1}{2}\|\mathbf{w}\|^2 + C\left(\sum_{i=1}^{n} \xi_i\right) \tag{2}$$

where C is the penalty factor, ξ_i is the slack variable.

At the same time, radial basis, i.e., gaussian kernel is selected as the kernel function of SVM training to obtain better detection accuracy due to the nonlinear problem of the actual situation of power system. The gaussian kernel function is shown in (3):

$$K(\mathbf{x}, \mathbf{x}_i) = \exp\left(-\frac{|\mathbf{x} - \mathbf{x}_i|^2}{\sigma^2}\right) \tag{3}$$

While parameters in Eq. (1) to (3) need to be fine-tuning according to the result of the simulation.

For islanding detection under different conditions, the core idea is that parameters should be fine-tuned according to specific conditions. Therefore, the selection of sampling points, feature vectors and data processing in the process of islanding is represented in Sect. 2.2.

2.2 Proposed Islanding Detection Method

The training accuracy of support vector machine is closely related to the feature vector input into SVM. Therefore, the thinking of data stack should be adopted to sample the electrical volume information at the sampling speed of 500 Hz and follow the stacking method to constantly update data to form data stack. During the training, the electrical volume sampling information of one cycle in front of the island and two cycles behind the island was selected for training, as shown in Fig. 1.

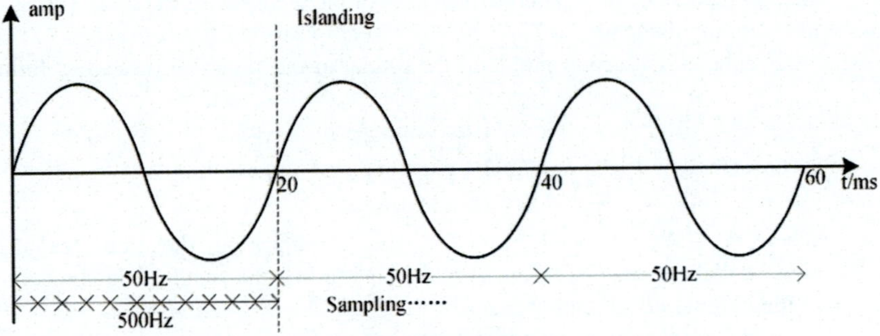

Fig. 1. Sampling diagram of electrical volume information.

The signal and sampling details are shown in Table 1. In this table, *RMS_U* is three-phase voltage root mean square of the PCC, *RMS_I* is three-phase current root mean square of the PCC, *P* and *Q* are the active power and passive power output of the PCC. Because of the symmetry of the islaing occasion, one phase value can satisfy the demand of detection. *f* is the frequency of PCC. Hence the detection flow can be took as below:

1) Construct a microgrid simulation model in real-time digital simulation(RTDS), the simulation results are obtained under different conditions whose details are shown in Sect. 3.
2) Sampling the electrical value of PCC in Table 1 and get 7 sampling matrices individually.
3) Label all the seven pieces of these samples as islanding(−1), grid disturbance(+1), considering the normal condition as one kind of grid disturbance.
4) Gain the feature vectors as formula (4). $v_1, v_2, \ldots\ldots, v_{30}$ are the feature vectors, i.e. the sampling of 7 electrical values.

$$T = [v_1, v_2, \ldots\ldots, v_{30}, y_i], \quad y_i \in \{-1, +1\} \tag{4}$$

5) Using LIBSVM [16] to train the SVM model and contrast the result.

Table 1. Signal diagram of electrical volume information

Signal	Sampling time(ms)	Amount of samples	Frequency(Hz)
RMS_U, *RMS_I*, *P Q*,*f*	2	30	500

3 Model Establishment and Simulation Results

3.1 Topology and Model Building in RTDS

In this paper, hydropower is selected as the main body of distributed power generation, which is equivalent to synchronous generator. The micro-grid formed by single power generation and load is selected for island detection. The connection topology of micro-grid and utility grid is shown in Fig. 2. The multi-DG in the figure are all low-capacity synchronous generators.

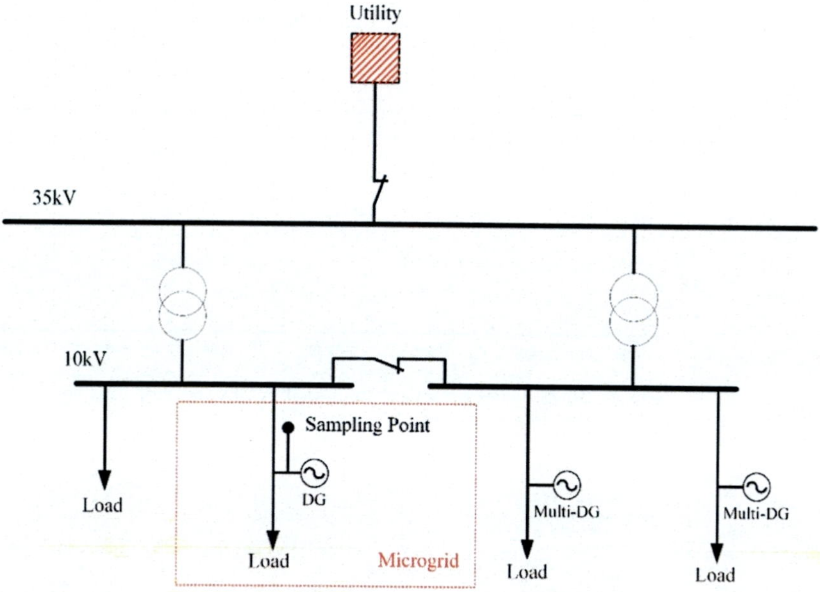

Fig.2. Topology of modeled system using RTDS.

According to this topology, the RTDS model is established, and run for 85 times in order to generate the data set. The 85 times islanding and non-islanding situation are shown in Table 2.

Table 2. Generated data set

Cases	Amount of simulate times	Description
Islanding	30	±30% active power mismatch
	15	Load Switching
	10	Other DG Switching
Non-islanding	15	Different Faults
	15	Light/Heavy load

3.2 Results of Simulation in RTDS

The results of simulation are divided into six different parts, i.e., three-phase voltage root mean square and values of voltage and current, frequency, active power output. According to various scenarios, namely, light load, ±30\% active power mismatch, second DG switching as the non-islanding occasion, DG switching, different kinds and location of faults, load switching as the islanding occasion. The comparison diagram is shown as Fig. 3, Fig. 4, Fig. 5, Fig. 6, Fig. 7 and Fig. 8.

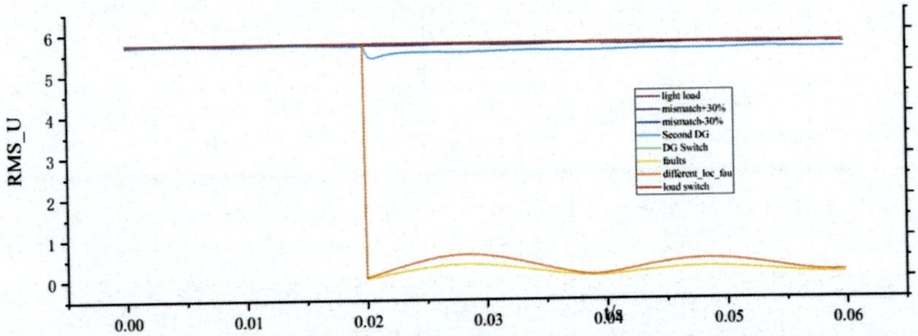

Fig. 3. RMS values of voltage in different cases and flow distribution.

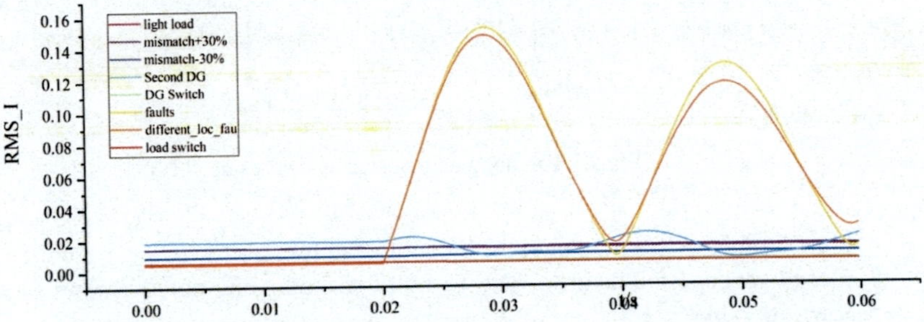

Fig. 4. RMS values of current in different cases and flow distribution.

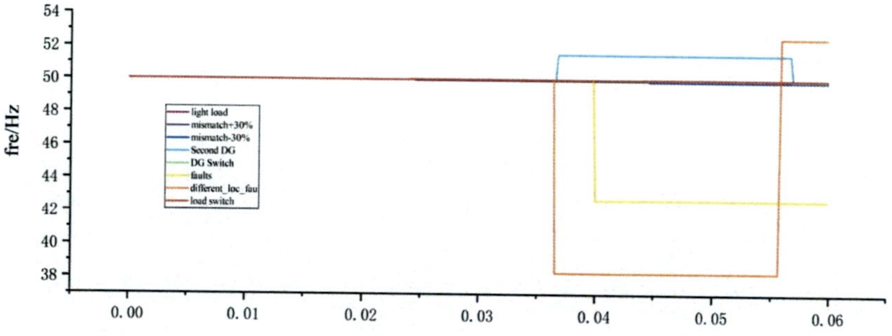

Fig. 5. Frequency in different cases and flow distribution.

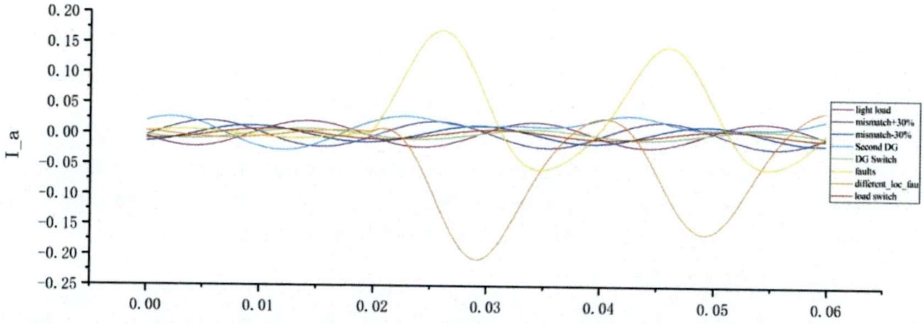

Fig. 6. Current of PCC in different cases and flow distribution.

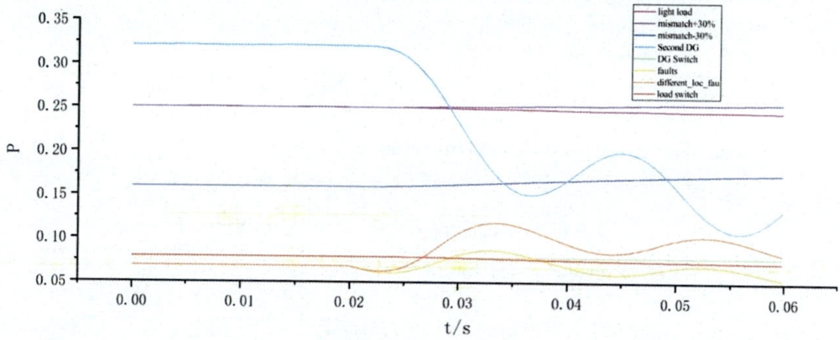

Fig. 7. Active power of PCC in different cases and flow distribution.

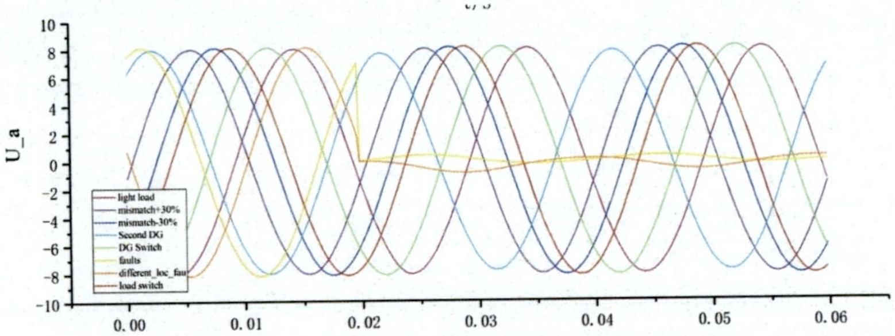

Fig. 8. Reactive power of PCC in different cases and flow distribution.

The simulation results were all selected to perform Islanding or non-Islanding switching operation after 0.02 s under the stable working state of the system to trigger the change of electrical volume and sampling with 0.002 s as the sampling period to obtain data and fit the curve.

It can be seen from the simulation results that the voltage has little change in non-Islanding and Islanding conditions, but the three-phase RMS of voltage has a significant difference, so does the current. The change of frequency has no clear rule, so it can only be used as a partial reference index. Therefore, only the three-phase root mean square of voltage and current and the active power output can be selected as the input vector of SVM for training.

3.3 Results Analysis of Support Vector Machine-Based Islanding Detection Method

As analyzed above, only three feature vectors, namely, three-phase root mean square of voltage, three-phase root mean square of current and active power output are used as input for SVM training.

Since the number of features is much larger than the number of samples, it is reasonable to use either gaussian kernel or linear kernel function for SVM training of samples. The results are shown in Table 3.

Table 3. Accuracy results of different kernel functions

Kernel function	Amount of test	Accuracy		
		RMS of Uabc	RMS of Iabc	P
Linear	30	100%	96.6%	96.6%
Guassian	30	93.3%	96.6%	100%
Polar	30	86%	83.3%	96.6%

Due to the small test set, this result cannot well explain the effectiveness of islanding detection by multiple SVMs, it can reach a high precision with the mixture of 3 well trained SVMs.

Since the results of machine learning may not be accurate, it is necessary to perform islanding detection again after a delay of 20ms. When the results of the two tests are consistent, the result is proved to be correct and can be output. But such an approach will lead to a new problem, that is, missed inspections in extreme cases.

4 Conclusion

This paper proposes an islanding detection method based on multi-SVM, three-phase root mean square of voltage and current, active power output are input as the feature vectors after analysing the waveform and data from the islanding and non-islanding occasion. According to the accuracy results of islanding prediction by using feature vectors above, it makes sense that linear kernel function gains a better performance, and active power output can be an important reference parameter during the detection process. Furthermore, the islanding detection based on support vector machine is recommended to be improved about the validation of its effectiveness because of the common dilemma, i.e., non-interpretability of machine learning.

Acknowledgment. This paper is supported by the Science and Technology Project of China Southern Power Grid (GDKJXM20200709).

References

1. Bakhshi-Jafarabadi, R., Sadeh, J., de Jesus Chavez, J., Popov, M.: Twolevel islanding detection method for grid-connected photovoltaic systembased microgrid with small non-detection zone. IEEE Trans. Smart Grid **12**(2), 1063–1072 (2021)
2. Baghaee, H.R., Mlakić, D., Nikolovski, S., Dragicević, T.: Support vector machine-based islanding and grid fault detection in active distribution networks IEEE. J. Emerg. Sel. Topics Power Electron. **8**(3), 2385–2403 (2020)
3. Matic-Cuka, B., Kezunovic, M.: Islanding detection for inverterbased distributed generation using support vector machine method. IEEE Trans. Smart Grid **5**(6), 2676–2686 (2014)
4. Alam, M.R., Muttaqi, K.M., Bouzerdoum, A.: An approach for assessing the effectiveness of multiple-feature-based SVM method for islanding detection of distributed generation. IEEE Trans. Ind. Appl. **50**(4), 2844–2852 (2014)
5. Ezzat, A., Elnaghi, B.E., Abdelsalam, A.A.: Microgrids islanding detection using Fourier transform and machine learning algorithm. Electric Power Syst. Res. **196**, 107224 (2021)
6. Mahela, O.P., Sharma, Y., Ali, S., Khan, B., Garg, A.R.: Voltage-based hybrid algorithm using parameter variations and Stockwell transform for islanding detection in utility grids. Informatics **8**(2), 21 (2021)
7. Mishra, S., Mallick, R.K., Gadanayak, D.A., Nayak, P.: A novel hybrid down sampling and optimized random forest approach for islanding detection and non-islanding power quality events classification in distributed generation integrated system. IET Renew. Power Gener. **15**(8), 1662–1677 (2021)
8. Ma, L., Guo, X., Wei, L.: An improved islanding detection algorithm based on AFDPF. In: E3S Web of Conferences, vol. 257 (2021)
9. Rabuzin, T., Hohn, F., Nordström, L.: Computation of sensitivity-based islanding detection parameters for synchronous generators. Electric Power Syst. Res. **190**, 106611 (2021)

10. Thakur, A.K., Singh, S.P., Shukla, D., Singh, S.K.: Passive method for islanding detection using variational mode decomposition. IET Renew. Power Gener. **14**(18), 3782–3791 (2020)
11. Eluri, N.V.D.V.P., Dash, P.K., Dhar, S.: Islanding detection in photovoltaic based DC micro grid using adaptive variational mode decomposition and detrended fluctuation analysis. IET Gener. Transm. Distrib. **15**(4), 631–644 (2020)
12. Abdi, H., Rostami, A., Rezaei, N.: A novel passive islanding detection scheme for synchronous-type DG using load angle and mechanical power parameters. Electric Power Syst. Res. (2020, prepublish)
13. Energy; Reports from University of Kentucky Describe Recent Advances in Energy: Islanding detection in rural distribution systems. Energy Wkly News (2020)
14. Energy - Renewable Energy; New Renewable Energy Findings Reported from National Institute of Technology Calicut: Synchrophasor based islanding detection for microgrids using moving window principal component analysis and extended mathematical morphology. J. Math. (2020)
15. Sathish, K.R., Ananthapadmanabha, T.: Islanding detection scheme of distributed generation systems using hybrid FAT-SGO approach. Int. J. Innov. Technol. Explor. Eng. (IJITEE) **10** (1) (2020)
16. Chang, C.-C., Lin, C.-J.: LIBSVM: a library for support vector machines. ACM Trans. Intell. Syst. Technol. **2**, 1–27 (2011). http://www.csie.ntu.edu.tw/cjlin/libsvm

Smooth Switching Strategy for Hydropower Microgrid with Auxiliary Load Shedding Control

Yuchen Huang[1], Zhicong Cheng[1], Li Li[1], Shan Lu[1], Jing Zhong[1], and Qiuping Zhang[2](✉)

[1] Guangdong Power Grid Co, CSG Shaoguan Power Supply Bureau, Shaoguan 510000, China

[2] School of Electrical Engineering and Automation, Wuhan University, Wuhan 430072, China

zhangqiuping@whu.edu.cn

Abstract. In remote mountainous areas with abundant hydropower resources, the power grid structure is single and the power supply reliability is low. When a blackout occurs at a higher level, the continuous power supply of important loads can be ensured through the smooth switching of hydropower microgrid. Therefore, in view of the failure of the upper power grid, this paper proposes a smooth switching optimization model for auxiliary load shedding control, which ensures the safety of the system and realizes more loads without stopping power. Since the proposed optimization model is a mixed integer programming model (MIP), the commercial solver that can solve the large-scale MILP model based on MATLAB can be used to obtain the optimization results.

Keywords: Smooth switching · Mixed integer programming model · Hydropower control

1 Instruction

With people's increasing demand for safety and economy of power grid, large-scale power grid is vulnerable to disturbance and poor quality of power supply in remote mountainous areas. Shaoguan area power grid structure is single, outage maintenance or line fault are easy to cause regional blackouts, power supply reliability is low. Due to the large number of rivers, sufficient rainfall, large terrain difference and rich water resources in the region, the total installed capacity of hydropower is huge, and some of the hydropower equipped with reservoirs have regulation ability. Therefore, the small hydropower can be connected with the power grid, and the hydropower microgrid with small hydropower as the main power source can be constructed according to local conditions to supply civil and industrial loads and improve the reliability of power supply.

Microgrid is a small generation and distribution system integrated by various distributed generators (DG), energy storage devices, load units, monitoring and protection devices and perfect energy management system, which can carry out self-control and self-energy management [1]. Different from wind, light and other new

Z. Hu et al. (Eds.): AIPE 2021, LNDECT 119, pp. 85–96, 2022.
https://doi.org/10.1007/978-3-030-97064-2_9

energy, small hydropower output has continuity and low volatility, which can be regarded as stable and controllable power supply. Taking small hydropower as the main power supply, its efficiency can be maximized to ensure power supply reliability and economy.

At present, there are many studies on the microgrid connected with new energy sources such as wind and light in China, and there are also some studies that consider the complementary characteristics of different power generation forms such as wind, light, water and storage [2, 3]. For example, Reference established the microgrid scheduling operation model with the optimal economic benefit as the goal to realize the efficient complementary operation of wind and water. Some scholars [4, 5] also studied the economic scheduling and optimal control of centralized small hydropower microgrid (group).

When a blackout occurs at a higher level, the planned islanding operation can be implemented, that is, to make a strategy in advance to transition to the islanding operation state. According to the time sequence, it can be divided into three processes: first, the main grid fault, the hydropower microgrid switch from the grid-connected state to the off-grid state; then, the hydropower microgrid maintains islanded operation mode; finally, the main grid is restored to failure, and the hydropower microgrid is restored to grid-connected operation mode from island operation mode. Due to the limited frequency modulation capacity of small hydropower, it is impossible to stabilize the huge unbalanced power caused by the loss of superior power supply, which is prone to frequency collapse. Based on this, this paper proposes an off-grid strategy of hydropower microgrid with auxiliary load shedding control to ensure continuous power supply of important loads.

In this paper, the smooth switching of the grid-connected and off-grid of the hydropower microgrid is studied. The existing research [6–11] improves the control mode of the smooth switching of the microgrid, and uses the mathematical programming method, heuristic algorithm or artificial intelligence algorithm to solve the problem. For example, in the literature [10], considering the reliability and power flow test, a dynamic island division model is established, and the frog leaping algorithm (SLFA) is used to solve the problem. Reference [11] A mixed integer linear programming model is established by integrating switching equipment, distributed generation and energy storage system, and multi-time step optimization is performed to form the final microgrid operation On this basis, the transient transition process is considered for smooth switching.

In terms of emergency frequency control, many references [12–14] conducted frequency dynamic analysis based on SFR model, and proposed various schemes for system frequency stability.

On this basis, combined with the frequency modulation model of hydropower, a mixed integer programming model is established, and the control strategy considering transient process is given, so as to complete the smooth switching of hydropower microgrid and ensure the continuous power supply of important load.

2 State Conversion of Hydroelectric Microgrid

The operation mode of hydropower microgrid is not single. According to its connection with the main network, it can be divided into two types, namely grid-connected mode and islanded mode. When the microgrid is in the grid-connected state and islanded state, the characteristic electrical quantity information is inconsistent, and the state identification and mode switching can be carried out accordingly.

2.1 Composition of Small Hydropower Microgrid

Small hydropower has small capacity, large quantity and distributed distribution. According to the analysis and summary of the way of small hydropower connecting to the grid in Shaoguan [15, 16], there are three typical structures of small hydropower connecting to the grid, as shown in the Fig. 1:

1) Series connection structure, the small hydropower directly connected to the 10 kV feeder, the formation of small hydropower nodes in turn access, namely the first line in the diagram 1;
2) Parallel structure, the small hydropower directly access to the 10 kV bus, forming a unified access node, namely the second line in the diagram 1;
3) Tree structure is a mixed mode of serial structure and parallel structure. The overall structure forms a T-shaped, complex and changeable, namely the third line in the diagram 1.

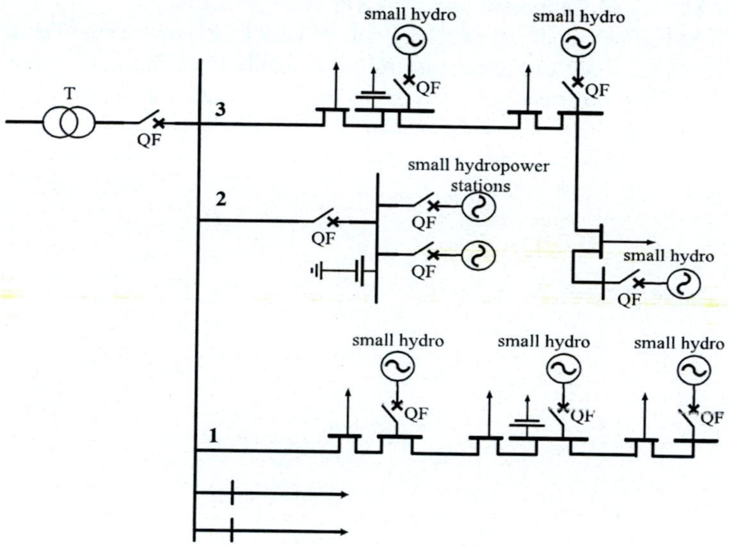

Fig. 1. Microgrid structure with only small hydropower

2.2 State Switching of Small Hydropower Microgrid

2.2.1 Grid-Connected and Off-Grid State Recognition

According to the rotor motion equation of the generator set, when the generator power is unbalanced, the generator speed will deviate from the rated speed, that is, the frequency will deviate from the rated frequency.

$$\begin{cases} \frac{d\delta}{dt} = \omega - \omega_0 \\ {}_J\frac{d\omega}{dt} = P_T - P_E - D(\omega - \omega_0) \end{cases} \tag{1}$$

where, ω is electrical angular velocity $(rad \bullet s^{-1})$, P_T is the mechanical power of prime mover, P_E is the electromagnetic power of generator, D is the damping, T_J is the inertia time constant of generator set.

According to the rotor motion equation of the generator set, under the normal operation of the large power grid, the fluctuation of the generator speed is small due to T_J is large. After being separated from the large power grid, it is small and the speed is easy to shake.

In the case of off-grid, there is a huge unbalanced power in the microgrid, and the frequency jitter is large. Due to the linear relationship between the frequency change rate and the unbalanced active power, the frequency change can be used as the off-grid identification criterion.

In the case of grid connection, the voltage and frequency are consistent with the large power grid, and are stable near the rated value for a long time. If the power disturbance is changed, the voltage and frequency are still stable near the rated value, then the microgrid is considered to be in the grid-connected state.

2.2.2 Grid-Connected and Off-Grid State Switching

Based on two modes of switching state of local hydropower controller device, namely automatic mode and remote control mode, mode switching can be completed. When the micro-grid controller sends the mode signal to the local controller, the local controller recognizes the grid-connected and off-grid state according to the recognition criterion while receiving the signal, and performs the action. The mode switching process is shown in the Fig. 2.

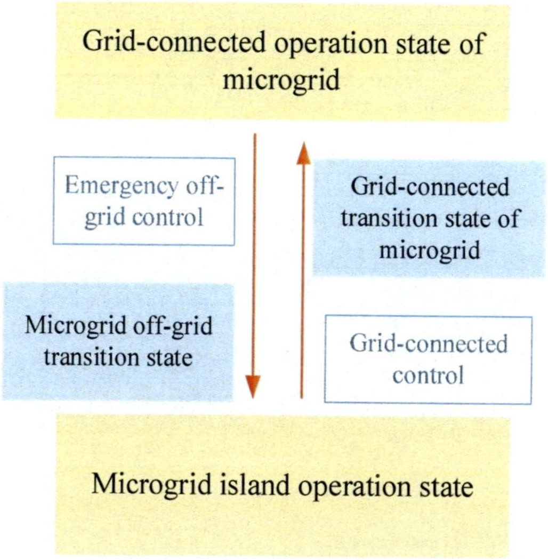

Fig. 2. Schematic diagram of mode switching

3 Control Model of Hydropower Units

When the hydropower microgrid is off-grid, there is a serious power shortage, that is, the hydropower output cannot meet the load demand, and the low frequency phenomenon will occur. Due to the limited hydropower regulation ability and regulation speed in short time scale, the auxiliary load shedding control enables the frequency to be stable in the safe operation range when the hydropower microgrid switches to the island mode.

3.1 Rational

According to the generator rotor motion equation shown by the formula, when the load fluctuates in the power grid, the turbine output should fluctuate to maintain the balance of mechanical power and electromagnetic power, and then control the generator speed stability, that is, the control frequency stability. The regulating system of hydraulic turbine includes control system and controlled system, which can carry out primary frequency regulation and secondary frequency regulation. Its structure is shown in the Fig. 3:

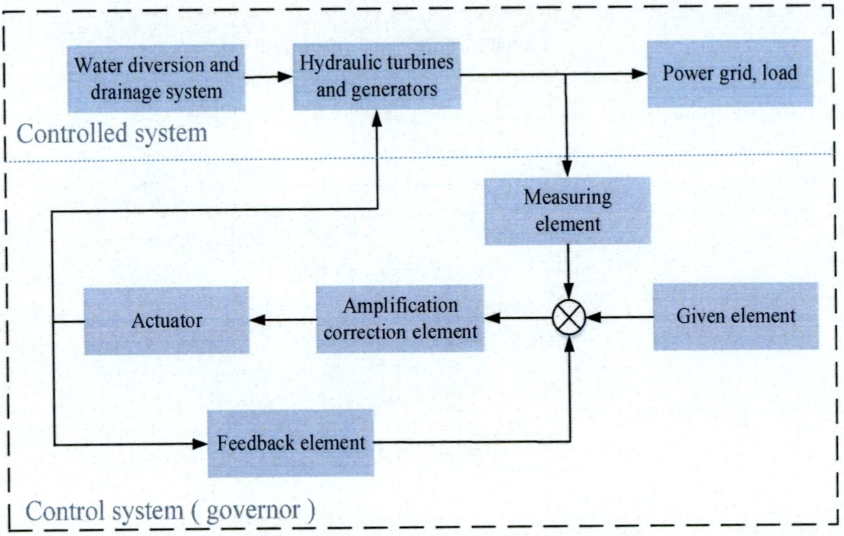

Fig. 3. Regulation system of hydraulic turbine

The control system of hydraulic turbine is mainly composed of governor and electro-hydraulic servo system. The former generally uses proportional-integral- differential regulation, and the latter is the actuator of main servomotor displacement. The above transfer function block diagram 4 is a complete logic block diagram of hydraulic turbine governing system, including guide valve, auxiliary relay, main pressure distribution valve, electro-hydraulic converter, main relay, secondary amplification and so on.

3.2 Control Model

Since the governor of small hydropower system controls the angle stroke directly by the servo motor, and the guidance of the turbine is directly controlled by the gear driven connecting rod to control the flow size without considering he permanent slip coefficient, the control system equation is implified as (2):

$$\frac{Y(s)}{\Delta F(s)} = K_P + \frac{K_D S}{1 + T_{DV} S} + \frac{K_I}{S} \tag{2}$$

where K_P is the proportional gain coefficient, K_D is the differential gain coefficient, K_I is the integral gain coefficient, and T_{DV} is the time constant of the differential link. These parameters can be calculated by the following parameters.

$$K_P = \frac{T_d + T_n}{b_t T_d},\ K_I = \frac{1}{b_t T_d},\ K_D = \frac{T_n}{b_t} \tag{3}$$

Among them, b_t is the transient difference coefficient, T_d is the delay time constant, and T_n is the acceleration time constant. These three parameters are jointly determined by the inertia time constant of the turbine unit T_w and the inertia time constant of the unit T_a.

$$\begin{aligned} &1.5\tfrac{T_w}{T_a} \leq b_t \leq 3\tfrac{T_w}{T_a} \\ &3T_w \leq T_d \leq 6T_w \\ &0.4T_w \leq T_n \leq 0.6T_w \end{aligned} \tag{4}$$

So far, the transfer function of the turbine set is constructed.

4 On-Line and Off-Line Switching Strategy of Hydropower Microgrid

When the hydropower microgrid is separated from the grid, it is necessary to reasonably divide the microgrid area by controlling the switch on and off according to the actual situation of the grid such as load power, hydropower capacity, hydropower regulation capacity and energy storage, so as to ensure the stable operation of the microgrid after it is separated from the main grid. Based on the engineering practice and the actual situation of Shaoguan area, it is a safe and stable off-grid mode to form a large microgrid after the off-grid of hydropower microgrid, which can quickly restore the grid connection after the main grid is stable.

4.1 Objective Function

The purpose of switching from grid-connected mode to off-grid mode is to ensure the power supply of important loads as much as possible, and on this basis, to ensure the power supply of more other loads as much as possible. Therefore, considering the importance of load, this paper introduces the weight coefficient to characterize the importance. In the actual model solution, in order to make the calculation results more intuitive, the three weights are reduced by 100 times, namely 1, 0.1, 0.01, as Table 1.

Table 1. Weight coefficient of Load

Importance of load	Primary load	Secondary load	Tertiary load
Weight coefficient λ_i	100	10	1

The off-grid objective is to minimize the cut load, and the objective function is:

$$F = \min \sum_{i=1}^{n} (1 - x_i) P_{Li} \lambda_i \tag{5}$$

In the formula (5), PLi is the load active power; x_i is a 0–1 variable, indicating the load state. If $x_i = 0$, it indicates the switch action, the load is removed; and if $x_i = 1$, it indicates the load is effective.

4.2 Constraint Condition

In the study of microgrid off-grid operation, constraints generally include topology constraints, power flow constraints, voltage constraints, frequency constraints, power constraints, equipment constraints, etc. Since this paper only considers the operation of a single microgrid after the hydropower microgrid is separated from the grid, the network topology constraints can be ignored. Without considering the load characteristics, it is considered that the load is a constant power load; on the basis of conventional steady-state constraints, considering the frequency modulation ability of small hydropower, the transient frequency constraint is added. The values used in this article are scalar.

(1) Power balance constraint, that is, the exchange power of the common contact point between the microgrid and the main network is 0.

$$S_{PCC} = 0 \tag{6}$$

(2) Voltage constraint, that is, the voltage of each node in the microgrid is within the upper and lower limits of the voltage.

$$\begin{gathered} V_{i\min} \le V_i \le V_{i\max} \\ V_{i\min} = 0.95, V_{i\max} = 1.05 \end{gathered} \tag{7}$$

(3) Frequency constraint, that is, the transient frequency and steady-state frequency are within the stable operation range.

$$\begin{aligned} & f_{t\min} \le f_t \le f_{t\max} \\ & f_{s\min} \le f_s \le f_{s\max} \\ & f_{t\min} = 0.99,\ f_{t\max} = 1.01 \\ & f_{s\min} = 0.996,\ f_{s\max} = 1.004 \end{aligned} \tag{8}$$

(4) Power constraint, that is, hydropower output does not exceed the limit.

$$\begin{gathered} P_{DGi\_\min} \le P_{DGi} \le P_{DGi\_\max} \\ P_{DGi\_\max} = 1.0, P_{DGi\_\min} = 0.2 \end{gathered} \tag{9}$$

where, S_{PCC} is the power of the public contact point; $V_{i\max}$, $V_{i\min}$ are the upper and lower limits of node voltage, $f_{t\max}$, $f_{t\min}$ are the upper and lower limits of transient frequency, $f_{s\max}$, $f_{s\min}$ are the upper and lower limits of steady frequency, and $P_{DGi\_\max}$, $P_{DGi\_\min}$ are the upper and lower limits of hydropower output, respectively.

4.3 Simulation Analysis

This paper only considers the microgrid with small hydropower as distributed generation, and builds a model with 6 units and 12 constant power loads on DigSILENT. The output power per unit of small hydropower is shown in the Table 2. Among them, there are three small hydropower stations with regulation ability, namely DG1, DG2 and DG3, corresponding to SM1, SM2 and SM3 in DigSILENT; other small hydropower does not have the ability to adjust; the unit power and weight coefficients of 12 loads are shown in the Table 3.

Table 2. Data of small hydropower

DG	Output active Power before off-grid	Upper limit of active power	Lower limit of active power	Maximum value of additional active power	Minimum value of additional active power
DG1	0.38	1	0.2	0.62	−0.18
DG2	0.95	1	0.2	0.05	−0.75
DG3	0.9	1	0.2	0.1	−0.7
DG4	0.65	1	0.2	0.35	−0.45
DG5	0.508	1	0.2	0.492	−0.308
DG6	0.56	1	0.2	0.44	−0.36

Table 3. Data of load

Load	PL1	PL2	PL3	PL4	PL5	PL6
Load power	0.28	0.32	0.28	0.3	0.8	0.9
Weight coefficient	0.01	0.1	1	0.1	1	1
Load	PL7	PL8	PL9	PL10	PL11	PL12
Load power	0.1	0.2	0.12	0.23	1.5	0.3
Weight coefficient	0.01	0.01	0.1	1	1	0.1

When the power shortage is 1.3, using MATLAB to solve the load shedding results are shown in the Table 4.

Table 4. The load shedding results

Load	PL1	PL2	PL3	PL4	PL5	PL6
Result	**0**	1	1	1	1	1
Load	PL7	PL8	PL9	PL10	PL11	PL12
Result	1	**0**	1	1	1	**0**

The results showed that the loads PL1, PL8 and PL12 were removed, and the primary load was not removed. The minimum transient frequency is 0.9934, and the steady-state frequency is 0.9979, which are within the threshold range, so that the frequency of the microgrid is stable after leaving the grid. On the basis of ensuring the continuous power supply of important loads, more loads are continuously uninterrupted.

Build the model in DigSILENT for validation. When the load is not removed, the result is shown in the Fig. 4; after the auxiliary load shedding control, the result is shown in the Fig. 5.

From the results, when the load is not removed, frequency will collapse. After the auxiliary load shedding control, the steady-state frequency is maintained at 0.997, which meets the threshold range of 0.996–1.004; the transient frequency is maintained at 0.9915, meeting the threshold range of 0.99–1.01, both the transient frequency and the steady frequency can operate within the range of security and stability. Therefore, the smooth Switching Strategy for hydropower microgrid with auxiliary load shedding control is effective.

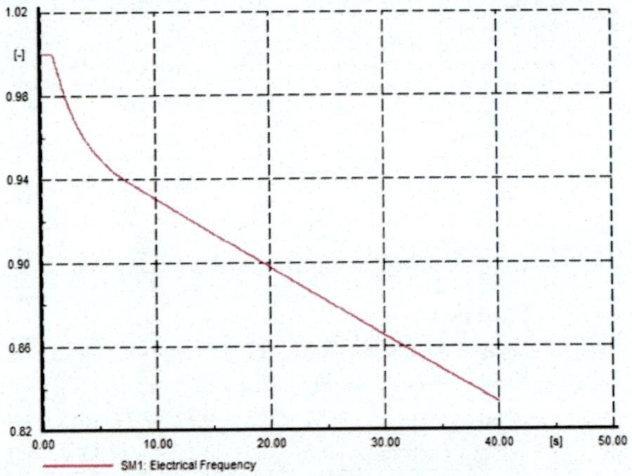

Fig. 4. Frequency curve before load shedding

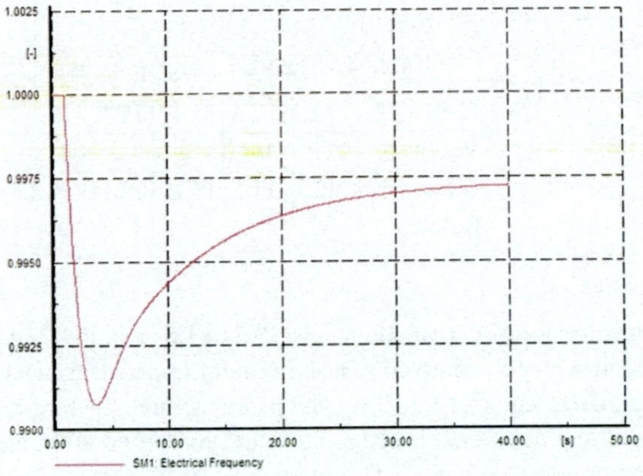

Fig. 5. Frequency curve after load shedding

5 Summary and Conclusion

Rich small hydropower is a unique advantage for the development of micro-grid in Shaoguang area. Hydropower resources are energy-saving, environmental protection and flexible dispatching, which are important energy guarantees for sustainable development. Aiming at the switching process of hydropower microgrid from grid-connected to off-grid, a mixed integer programming model with the minimum load shedding as the goal and the steady-state constraints and transient constraints as the constraints is established. The smooth switching strategy of microgrid is formulated, and the effectiveness of the strategy is verified based on MATLAB and DigSILENT. The analysis shows that this strategy has certain practical significance for realizing the balanced off-grid of hydropower microgrid, ensuring the continuous power supply of load and ensuring the safe and stable operation of hydropower microgrid.

Acknowledgment. This project is supported by Key Projects of China Southern Power Gird (GZ2014-2-0049).

References

1. Chen, Z., Wang, K., Li, Z., Zheng, T.: A review on control strategies of AC/DC micro grid. In: 2017 IEEE International Conference on Environment and Electrical Engineering and 2017 IEEE Industrial and Commercial Power Systems Europe (EEEIC/I CPS Europe), pp. 1–6 (2017)
2. Liu, Y.-H., Zhang, N., Zhang, X.: Research on grid-connected/islanding smooth switching of micro-grid based on energy storage. In: 2012 IEEE International Conference on Power System Technology (POWERCON), pp. 1–5 (2012)
3. Aponte-Roa, A., Cabarcas, G.D.G., Weaver, W.W.: AC vs DC power efficiency comparison of a hybrid wind/solar microgrid. In: 2020 IEEE Conference on Technologies for Sustainability (SusTech), pp. 1–5 (2020)
4. Guan, Y., Vasquez, J.C., Guerrero, J.M., Wu, D., Feng, W., Wang, Y.: Frequency stability of hierarchically controlled hybrid photovoltaic-battery-hydropower microgrids. In: 2014 IEEE Energy Conversion Congress and Exposition (ECCE), pp. 1573–1580 (2014)
5. Cai, C., Cheng, S., Jiang, B., Dai, W., Wu, M., Zhang, J.: Optimal operation of microgrid composed of small hydropower and photovoltaic generation with energy storage based on multiple scenarios technique. In: 2015 5th International Conference on Electric Utility Deregulation and Restructuring and Power Technologies (DRPT), pp. 2185–2190 (2015)
6. Zhu, J., Gu, W., Jiang, P., Wu, Z., Yuan, X., Nie, Y.: Integrated approach for optimal island partition and power dispatch. J. Mod. Power Syst. Clean Energy **6**(3), 449–462 (2017). https://doi.org/10.1007/s40565-017-0314-z
7. Popovic, Z.N., Knezevic, S.D., Brbaklic, B.S.: A risk management procedure for island partitioning of automated radial distribution networks with distributed generators. IEEE Trans. Power Syst. **35**(5), 3895–3905 (2020)
8. Rosero, C.X., Velasco, M., Martí, P., Camacho, A., Miret, J., Castilla, M.: Analysis of consensus-based islanded microgrids subject to unexpected electrical and communication partitions. IEEE Trans. Smart Grid **10**(5), 5125–5135 (2019)

9. Kamali, S., Amraee, T., Fotuhi-Firuzabad, M.: Controlled islanding for enhancing grid resilience against power system blackout. IEEE Trans. Power Deliv. **36**(4), 2386–2396 (2021)
10. Chen, B., Chen, C., Wang, J.: Multi-time step service restoration for advanced distribution systems and microgrids. IEEE Trans. Smart Grid **6**(9), 6793–6805 (2018)
11. Oboudi, M.H., Hooshmand, R., Karamad, A.: Feasible method for making controlled intentional islanding of microgrids based on the modified shuffled frog leap algorithm. Int. J. Electr. Power Energy Syst. **78**, 745–754 (2016)
12. Smolovik, S.V., Koshcheev, L.A., Lisitsyn, A.A., Denisenko, A.I.: Special automation for isolated power systems emergency control. In: 2021 IEEE Conference of Russian Young Researchers in Electrical and Electronic Engineering (ElConRus), pp. 1558–1561 (2021)
13. Xue, A., Zhang, J., Zhang, L., Sun, Y., Cui, J., Wang, J.: Transient frequency stability emergency control for the power system interconnected with offshore wind power through VSC-HVDC. IEEE Access **8**, 53133–53140 (2020)
14. Gurov, N., Chuvychin, V., Andreevsky, I., Vorshev, A.: Problems of power system frequency and active power control. In: PowerTech Budapest 99. Abstract Records. (Cat. No. 99EX376), p. 7 (1999)
15. Ye, L., Zhang, Y., Xia, C., Hu, W., Chen, L.: The research and application of the optimal operation based on improved genetic algorithm for small hydropower stations in Shaoguan. In: 2008 International Conference on Electrical Machines and Systems, pp. 2707–2711 (2008)
16. Yongjin, C., Zhifeng, C., Yanchun, X., Junxiong, L.: Analysis of operation characteristics of small hydropower isolated network during leading phase. In: 2020 Asia Energy and Electrical Engineering Symposium (AEEES), pp. 859–862 (2020)

Automatic Identification Technology for Distribution Terminals Based on Unlicensed LPWAN

Ran Hu[1], Xutao Shi[2], Lei Yu[2], Zhiyong Yuan[2], Zhanhua Huang[1], Kairan Li[2], and Gaomin Zhang[3(✉)]

[1] Shenzhen of Guangdong Power Grid Co., Ltd, Shenzhen 518000, China
[2] Electric Power Research Institute, CSG, Guangzhou 510000, China
[3] School of Electrical Engineering, Wuhan University, Wuhan 430072, China
2015302540163@whu.edu.cn

Abstract. In order to realize the plug and play of the power distribution terminal, the distribution network communication is divided into three parts: the manager, the distribution terminal and the communication network. This article is based on an unlicensed frequency band LPWAN technology, i.e. LoRa, through the discovery, return, analysis, update and control process to complete the automatic identification and plug-and-play of the power distribution terminal equipment information. First, the master station center (manager) sends Discover information to the power distribution terminal through the communication network, and then the power distribution terminal returns the equipment information, the manager analyzes and updates the master station model, and finally completes the automatic identification.

Keywords: Distribution communication · Distribution terminals · Automatic identification · Unlicensed spectrum · LoRa

1 Introduction

Since smart grid has been developed in China for many years, the level of automation and demand for power distribution systems have continued to increase. At the same time, the communication capabilities of the power distribution links of the power system have been continuously improved, and communication methods have become more diversified. Therefore, with the access of massive distribution network smart terminals, the current power distribution link urgently needs to improve the plug-and-play and automatic identification technology of smart terminals to improve work efficiency, reduce the workload of the construction, maintenance and debugging of the distribution network automation system, and further Strengthen the monitoring efficiency and power supply reliability of the electrical energy indicators of the distribution network, and improve the automation level of remote measurement, remote signaling and remote adjustment of the distribution network.

However, in the current distribution network, the distribution intelligent terminal and the distribution network master station usually complete the debugging through the

Z. Hu et al. (Eds.): AIPE 2021, LNDECT 119, pp. 97–103, 2022.
https://doi.org/10.1007/978-3-030-97064-2_10

cable or the power dedicated network based on the information forwarding table. The debugging process usually relies on manual point-by-point verification. Operation errors or other human factors can cause problems such as incomplete information, errors, and weak relevance of power distribution terminals. The communication system of the traditional power distribution network IEC61850 digital power station is supported by the IP communication network to provide the realization basis for the automatic identification and plug-and-play of the equipment. Therefore, many researches are carried out on the plug-and-play of the distribution network.

Reference [1] research on the plug-and-play technology of intelligent terminal equipment management in the distribution network based on the idea of the Internet of Things, to realize the automatic identification of the intelligent terminal equipment in the distribution network. Reference [2] focuses on the research of key plug-and-play technologies for the user's quantum keys in the time domain and frequency domain. Reference [3] completed the research on automatic identification of distribution network equipment with distributed energy based on the IEC61850 communication system, which further promoted the establishment of a plug-and-play system for smart terminals. Reference [4] realizes the distributed energy energy management system of smart grid based on automatic identification of equipment, while Reference [5] implements relay protection based on plug-and-play technology in the distribution network scenario with distributed power generation, It is of great significance to the top-level application of automatic device identification. Reference [6–8] Research on the Plug and Play Quantum Key Distribution Method.

However, the above automatic identification methods are all based on the existing IP communication network. With the development of the communication network, the unlicensed spectrum low power wide area network, LPWAN is widely sought after for its excellent market prospects and good communication performance. LoRa, as the LPWAN technology of unlicensed spectrum, is suitable for the implementation of plug-and-play technology in the distribution network. This article is based on the unlicensed spectrum LPWAN, namely LoRa, through the discovery, return, analysis, update and control process to complete the plug-and-play of intelligent power distribution terminals Ready to use and automatic recognition.

2 Low Power Wide Area Network

2.1 Unlicensed Spectrum Low Power Wide Area Network

LPWAN technology is divided into licensed spectrum and unlicensed spectrum. Licensed spectrum refers to the radio frequency band that needs to be applied for, and unlicensed spectrum LPWAN refers to spectrum that can be used free of charge without application. Unlicensed LPWAN mainly includes LoRa, Sigfox and NWave, etc. The literature [9] compares unlicensed spectrum communication technologies. This article combines power distribution scenarios and summarizes it as shown in Table 1.

Table 1. Comparison of main technical characteristics of unlicensed spectrum LPWAN in power distribution scenarios

	LoRa	Sigfox	NWave	Telensa
Distance/km	15–45(Rural) 3–8(Urban)	40(Rural) 10(Urban)	10	Ceiling 8
Frequency/Mhz	470–510	868, 902	Sub-Ghz	470(China) 868,975
ISM spectrum	Yes			
Data Rate	0.3 kbps–50 kbps	100 kbps	100 kbps	Low Rate
Distribution Operation Mode	PowerGridCo./user	PowerGridCo	PowerGridCo./user	PowerGridCo
Economic Index	High			
Standardized	LoRaWAN	–	Weightless-N	–
Data mining	PowerGrid	Authorization By Sigfox	Non-public	Non-public

Unlike 4G, 5G and other authorized spectrum communication technologies, the advantages of unlicensed spectrum communication technology in power distribution networks are low-cost operation mode, anti-interference ability, and a good ecosystem. In terms of communication technology transmission distance, Telensa alone cannot meet the terminal information transmission requirements required by urban and rural power distribution systems. At the same time, the topological data connection of the distribution network also comes from the intelligent terminals distributed throughout the distribution network, so the data ownership of the communication technology should be able to be delivered to the grid company.

In terms of communication data, the main communication data of the terminal includes protection, measurement and control, and power information. The above-mentioned LPWAN technology meets the requirements of the data communication rate of the power distribution network. In summary, LoRa is a relatively suitable technology for terminal communication among unlicensed spectrum communication technologies.

2.2 LoRa in the Distribution Network

Since LoRa communication was proposed, there has been good research and development in terms of communication distance, cost, power consumption, speed, etc. [10–13]. It has long distance, low power consumption, low cost, easy deployment, and standardization. Perfect and other advantages, its components in the distribution network should include the following three parts:

1) Intelligent terminal: i.e. the realization of the physical layer, MAC layer and application layer, directly interact with voltage transformers and current transformers, and perform point-to-point transmission to the power distribution master station through LoRa standardized protocol specifications.

2) LoRa Gateway: Receiving upstream data, solving data concurrency problems, and being close to the installation location of power distribution terminals, providing two way communication capabilities for smart power distribution terminals.
3) Power distribution master station: mainly contains a variety of servers, responsible for terminal and gateway management of LoRa communication, analysis and processing of electrical data, human-computer interaction, etc.

The main table parameters of LoRa communication include Band-width (BW), Spreading Factor (SF) and Code Rate (CR). For symbol period Ts and symbol rate Rs satisfies formula (1) [11]:

$$\begin{cases} T_s = \frac{2^{SF}}{BW} \\ R_s = \frac{1}{T_s} = \frac{BW}{2^{SF}} \end{cases} \tag{1}$$

The code rate CR of LoRa satisfies formula (2):

$$CR = \frac{n}{n+4}, \quad n \in \{1, 2, 3, 4\} \tag{2}$$

Return formula (2) to formula (1) to get the bit rate R_b:

$$R_b = SF \cdot \left(\frac{BW}{2^{SF}}\right) \cdot CR \tag{3}$$

For power systems, alternating magnetic fields caused by three-phase alternating current, and broadband spectrum noise superimposed on phase lines, center lines or signal lines are very common, especially in the process of cable communication and carrier communication in the distribution network. Control factors may lead to ambiguous communication problems. Therefore, the CR value of LoRa communication needs to be adaptively reduced to reduce the packet error rate caused by Gaussian noise type interference to meet the anti-interference requirements of data transmission of intelligent power distribution terminals.

LoRa communication technology includes three working modes. This article believes that according to the requirements of LoRa communication technology, the distribution network intelligent terminal should be a Class A working mode, i.e. two-way terminal equipment. In the normal working environment of the power distribution network, the terminal usually only needs data transmission of remote measurement, remote signaling, and remote adjustment. Therefore, scheduling based on the random time base can be performed based on the actual situation of data transmission to save energy consumption. In the process of installation and configuration of intelligent terminals such as feedback terminal unit (FTU), distribution terminal unit (DTU), Transformer terminal unit (TTU), etc., it is necessary to transmit the registration information of the equipment to the distribution master station, ie manager, based on LoRa communication, Carrying out information interaction with the manager, building a device database, and forming a plug-and-play system after automatic identification.

3 Automatic Identification of Power Distribution Terminal

The automatic identification of power distribution terminals in the power system usually adopts object-oriented modeling ideas [15]. The advantage of this method is that the terminal information of the equipment can be hierarchically grouped, and multiple data attributes can be divided according to their functions. The terminal identification of the electric master station provides the basis.

3.1 Information Model Establishment

If the information model is established in an object-oriented manner, the actual situation of the distribution network should be considered for the corresponding information model layering. Therefore, this paper divides the information model into four modules, namely the measurement and control function module, the protection function module, and the power supply. Management module and device information module. The measurement and control function module is mainly responsible for transmitting the switch position information, harmonic information, phase sequence and unbalance, current and voltage of the distribution network back to the main station, and the protection function module is mainly responsible for the overcurrent protection, ground fault protection and protection of the distribution network. Location information records, the power management module is mainly responsible for providing battery status data of the terminal board, and the device information module mainly contains the configuration, description and extended information of the device. Build the information model together with four modules.

3.2 Automatic Identification of Terminal Equipment

The automatic identification of the intelligent terminal needs to be connected with the communication technology to ensure the correctness of the logical relationship. Therefore, the automatic identification process of the discovery registration model based on LoRa communication technology is proposed, as shown in Fig. 1.

When the power distribution terminal is installed and debugged, the master station issues Discover information, and the power distribution terminal returns Register information. If the terminal does not receive a downlink response, it will check whether the limit delay time in LoRa communication meets the requirements, and then check again. Whether the data transmission schedules the lowest rate in the strategy table to determine the success of LoRa communication. If the communication is successful, compare the terminal information and the master station information to update, and complete the automatic identification and plug-and-play of the equipment.

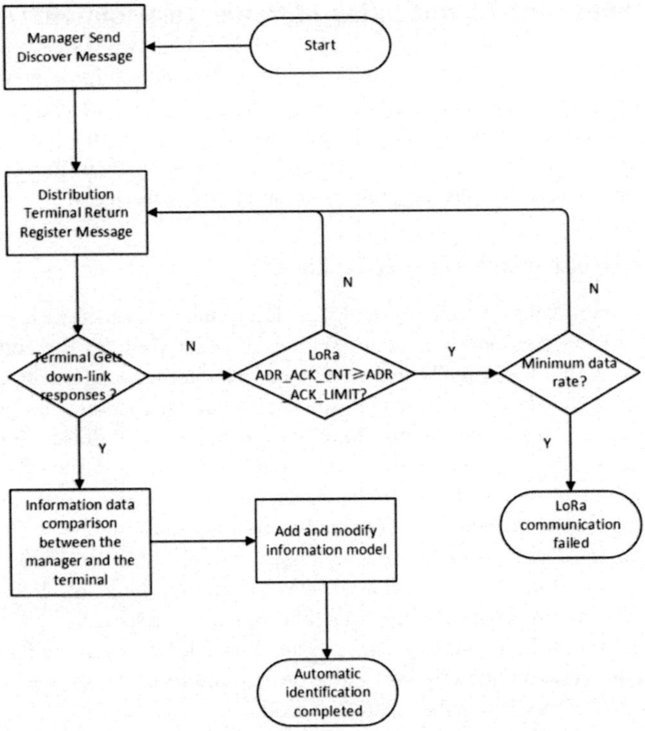

Fig. 1. Flow chart of automatic identification of power distribution terminal based on LoRa.

4 Conclusion

This paper proposes a plug-and-play and automatic identification method for power distribution terminals based on LoRa communication technology. It is limited by equipment and cannot be verified by experiments. In theory, the communication capabilities of LoRa technology can help power distribution terminals to access quickly Accurately update the equipment information of the power distribution terminal from the model library of the master station to realize the automatic identification function of the power distribution terminal.

Acknowledgment. This paper is supported by the Science and Technology Project of China Southern Power Grid (SZKJXM20200434). At the same time, the authors would like to thank Dr. Lihan for his reference opinion with this paper.

References

1. Lin, J., Wang, P., Zhang, J., Zhang, Z., Sun, H.: Plug and play technology for power distribution terminal management based on the IoT ideas. In: 2019 IEEE 8th International Conference on Advanced Power System Automation and Protection (APAP), pp. 196–200 (2019)
2. Aladadi, Y.T., Abas, A.F., Alwarafy, A., Alresheedi, M.T.: Multiuser frequency-time coded quantum key distribution network using a plug-and-play system. In: 2018 International Conference on Optical Network Design and Modeling (ONDM), pp. 53–58 (2018)
3. Ahn, B., Ha, J., Seo, Y., Heo, J., Shin, J., Lee, K.: Implementation of plug and play quantum key distribution protocol. In: 2018 Tenth International Conference on Ubiquitous and Future Networks (ICUFN), pp. 47–49 (2018)
4. Pan, S., Liu, D., Li, Q., Nong, J.: Research on plug-and-play of der-network coordinate controller based on IEC 61850. In: International Conference on Renewable Power Generation (RPG 2015), pp. 1–6 (2015)
5. Zhang, Z., Zhang, Y., Chow, M.-Y.: Distributed energy management under smart grid plug-and-play operations. In: 2013 IEEE Power Energy Society General Meeting, pp. 1–5 (2013)
6. Lydon, B.: BioPhorum plugfest tests plug-and- play equipment interoperability. InTech **68** (3) (2021)
7. Qi, R., Zhang, H., Gao, J., Yin, L., Long, G.-L.: Loophole-free plug-and-play quantum key distribution. New J. Phys. **23**(6) (2021)
8. Wu, X., Wang, Y., Huang, D., Guo, Y.: Multi-mode plug-and-play dual-phase-modulated continuous-variable quantum key distribution. Quantum Inf. Process. **20**(4) (2021)
9. Tsimtsios, A.M., Nikolaidis, V.C.: Towards plug-and-play protection for meshed distribution systems with DG. IEEE Trans. Smart Grid **11**(3), 1980–1995 (2020)
10. Su, G.-T., Zhao, J.: Analysis on LoRa wireless network technology. Mob. Commun. **40**(21), 50–57 (2016)
11. Edward, P., El-Aasser, M., Ashour, M., Elshabrawy, T.: Interleaved chirp spreading LoRa as a parallel network to enhance LoRa capacity. IEEE Internet Things J. **8**(5), 3864–3874 (2021)
12. Croce, D., Gucciardo, M., Mangione, S., Santaromita, G., Tinnirello, I.: Impact of LoRa imperfect orthogonality: analysis of link-level performance. IEEE Commun. Lett. **22**(4), 796–799 (2018)
13. Elshabrawy, T., Robert, J.: Interleaved chirp spreading LoRa-based modulation. IEEE Internet Things J. **6**(2), 3855–3863 (2019)
14. Wenyan, Z.: Technical overview on LoRa physical layer and MAC layer. Mob. Commun. **41** (17), 66–72 (2017)
15. Wang, Y., Xu, X., Mei, J., et al.: Automatic recognition technology of distribution terminal based on IEC 61850. Electr. Meas. Instrum. **53**(6), 32–37 (2016)

Multi-machine Joint Video Tracking Mechanism and Its Application for Substation Safety Protection

Tao Qian(✉), Hao Sun, Bing Han, Shuai Zou, and Fangwei Zhong

Xinjiang Information Industry Co., Ltd., Urumqi 830000, Xinjiang, China

Abstract. With the improvement of my country's industrialization level and the development of network information technology, automation technology has deeply served various industrial fields. Therefore, this paper uses the five-frame difference method in the inter-frame difference method in combination with the actual operation requirements of the substation, which is more traditional. The standard three-frame difference can better eliminate the hole; then the moving target is tracked, and the mean-shift-based cam-shift image tracking technology is used. Finally, through MATLAB under the WINDOWS system, the program operation verification and analysis of the video data of the substation is carried out. The results show that the method in this paper can better meet the requirements of the video monitoring of the substation.

Keywords: Substation · Security protection · Video · Tracking · Mechanism · Application

1 Instruction

At present, the intelligent video surveillance system [1, 2] has made good progress. Information communication and image transmission technology have become perfect, and have begun to emerge in various fields. For the safety protection of substations that pay more attention to safety and real-time, Intelligent video surveillance system is very important. Compared with the traditional substation safety protection system, intelligent video surveillance has incomparable advantages.

Intelligent video surveillance system has been studied in Western countries for a long time. It was first seen in the remote video surveillance system jointly developed by US Geological Survery's Center for Coastal Geology and Oregon State University in 1991. This system is mainly used to monitor the changing trend of coastal terrain. Image recognition is used to diagnose and analyze coastal changes in different periods. In addition, in 1997 DSRPA funded more than a dozen universities and research institutes such as Carnegie Mellon University and Massachusetts Institute of Technology to jointly develop VSAM (Visual Surveillance and Monitoring) [3], which is used in multi-object surveillance in battlefields and civilians. Achieved good results. The W4 [4] system developed by the University of Marangli is also quite successful in

Z. Hu et al. (Eds.): AIPE 2021, LNDECT 119, pp. 104–113, 2022.
https://doi.org/10.1007/978-3-030-97064-2_11

judging whether a person is carrying something. In the tracking and monitoring of vehicles and people, the University of Reading started earlier. At present, the three top computer vision conferences in the world: IEEE Conference on Computer Vision and Pattern Recognition, IEEE International Conference on Computer Vision, and European Conference on Computer Vision also list intelligent video surveillance as their main topics. In terms of commercial use, well-known companies such as IBM and Microsoft are applying intelligent video surveillance systems to commercial applications. At present, the more well-known companies in the commercial field include Object Video from the United States, IOImage from Israel, DETEC AC from Norway, and Axis from Sweden.

Compared with foreign countries, domestic research on intelligent video surveillance systems started relatively late, but developed rapidly. The Automation Institute of the Chinese Academy of Sciences is in a leading position in China. In addition, a series of university research institutes and national key laboratories such as Peking University and Tsinghua University [5] have also achieved good results. The National Intelligent Vision Surveillance Conference has held four sessions from 2002 to 2016 from time to time, providing an opportunity for national talents studying intelligent vision to communicate with each other and making outstanding contributions to the field of intelligent vision in my country. There are also many domestic companies that have started the research on intelligent video surveillance and have all been applied in practice [6]. For example, Shenzhen Huirui Tianyan Company has been certified as a national "high-tech enterprise", Suzhou Huayi Security Technology Co., Ltd., Shenzhen Beixin Intelligent System Co., Ltd. and so on. The intelligent monitoring systems of these companies have been applied in petrochemicals, road violations, hotels, villas, shopping malls, gymnasiums and other fields.

This article is based on mean-shift cam-shift, and uses cam-shift to track moving targets. Finally, this article applies the introduced method and algorithm to actual production, and basically solves a series of practical problems such as safety helmet detection.

2 Detection of Moving Targets

Moving target detection [7–9] is also called moving target segmentation, which is a technique that separates the moving target from the stationary background, and only obtains the moving target we need while removing the background. Moving target detection can be divided into static background and dynamic background according to whether the background changes. This article only considers the moving target detection method under static background. In addition, in a static background, due to the influence of factors such as light and dark changes, vibration, etc., it will cause certain difficulties in the detection of moving targets.

This section first introduces the three commonly used detection methods in detail, compares and analyzes their respective advantages and disadvantages one by one, and then chooses the inter-frame difference method as the moving target detection method in this article in accordance with the actual needs of the substation. The essence of moving target detection is to separate the moving target from the background image. After segmentation, in order to make the target more prominent, the image needs to be binarized. In the binarization process, how to choose the correct threshold [10] is the key.

For some simple images, the gray levels of the background and the moving target are very different, that is, the contrast is very large. When displayed in the histogram, there are two obvious peaks on the left and right. At this time, the trough is selected as the threshold to compare precise. But for complex images, there are no conspicuous peaks and valleys in the histogram, so this method is naturally not applicable.

In addition to the above-mentioned bimodal method, the commonly used threshold selection methods include P parameter method, maximum between-class variance method (Ostu, Otsu method), maximum entropy threshold method, and iteration method (optimum threshold method). This article chooses the iterative method [11], which continuously divides the image into regions, and selects the thresholds of the regions respectively, and continuously loops until the optimal threshold is selected. The specific steps are:

(1) Calculate the maximum Z_{max} and minimum Z_{min} of the image gray value, and use the average gray value as the initial threshold T_k, $T_k = \frac{Z_{max} + Z_{min}}{2}$;
(2) Continue to calculate the average gray values Z_A and Z_B for the two divided regions:

$$\begin{cases} Z_A = \frac{\sum_{Z(i,j) < T_k} Z(i,j)}{N} \\ Z_B = \frac{\sum_{Z(i,j) > T_k} Z(i,j)}{M} \end{cases} \tag{1}$$

Among them, $Z(i,j)$ is the gray value of pixel (i,j), and N is the number of pixel (i,j) when $Z(i,j) < T_k$,
M is the number of pixels (i,j) when $Z(i,j) > T_k$.
(3) Calculate the new threshold $T_{k+1} = \frac{Z_A + Z_B}{2}$;
(4) End the loop when $T_k = T_{k+1}$, otherwise $T_k = T_{k+1}$, go to step 2.

3 Moving Target Tracking

3.1 Moving Target Tracking Method

Region-based matching tracking method. The basic principle of this method is to extract the features of the connected area of the original moving target, then calculate the area in the current frame image that is roughly the same as the stored original moving target area, and then track the area. Obviously, when the moving target is not occluded, the accuracy and stability of this method are relatively satisfactory. However, this method requires a large amount of calculation, especially when the area to be tracked is large, it takes a long time to calculate the features, which is not conducive to real-time tracking. In addition, this method requires that the moving target cannot have too much deformation. For example, the target becomes larger and lower due to the distance from the camera, which will affect the tracking. Excessive occlusion of the moving target will also reduce the accuracy and cause the target to be lost. Therefore, Jorge et al. proposed an improvement to the method, which combines moving target tracking with moving target detection and segmentation. The target segmentation is real-time, the results of the real-time segmentation are feature extraction, and the matching area is updated in real time. At the same time, the tracking information is fed back to the target detection to complement each other.

Tracking based on feature points (key points). KLT is a feature point tracking algorithm that is widely used among them. It relies on the image sequence generally has a relatively small time interval. The feature points of the moving target are generally smooth in the continuous image sequence, so it can be smoothed according to the feature. To perform target tracking. This algorithm has a better tracking effect when the target is partially occluded, because the feature of the target is generally not single. When the target is partially occluded, other features can be selected to continue tracking. For example, when two pedestrians occlude each other, although some characteristics disappear, the two pedestrians can still be better tracked according to the difference in the speed of the center of mass of the two pedestrians. The difficulty of this algorithm lies in how to accurately complete a series of tasks such as acquiring, transforming, storing, and clearing feature points when the target is occluded.

Tracking based on geometry. This method has relatively large limitations and is only suitable for the tracking of rigid objects, such as the tracking of cars. For non-rigid targets, because the targets often have various irregular deformations, it is difficult to obtain an accurate geometric structure model and cannot be achieved. track.

In addition, there is the Snake model based on active contours, which is particularly suitable for tracking deformed targets. This algorithm is often combined with Kalman filtering, and the effect is more ideal, but it is generally used for tracking a single target, and multi-target tracking generally uses an active contour model based on a level set. There is also the optical flow method which is still in the theoretical stage. At present, the mean-shift tracking algorithm is the most widely used [12].

3.2 Cam-Shift Tracking Algorithm

The full name of cam-shift is Continuously Adaptive Mean Shift Algorithm, which is an improvement of the mean-shift algorithm. The mean-shift tracking algorithm has artificially selected a detection window in the first step. The size of this window is unchanged. When the tracking target becomes smaller, the useless background information contained in the window will increase, which will cause the accuracy is reduced, and in actual applications, tracking may fail due to changes in the distance between the target and the camera. If mean-shift is the operation tracking of a frame of image, then cam-shift is the tracking of a series of video frames. Cam-shift is to perform the mean-shift algorithm operation on a series of video frames first, and apply the size and centroid of the detection window calculated in the previous frame to the processing of the next frame image, so that the window changes with the tracking target Yes, the tracking results are more accurate, and the tracking will not fail due to the distance between the tracking target and the camera.

The existing image color models are basically RGB models, which is what we call the principle of three primary colors. In fact, there are not only three colors in nature. In fact, each color has its own color spectrum. The colors can be regarded as the combination of three primary colors according to different percentages. Because it is three-dimensional, it is easier to understand, as shown in Fig. 1:

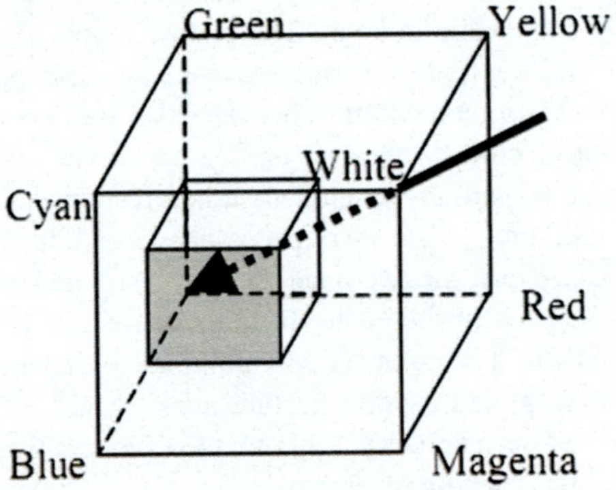

Fig. 1. RGB color model

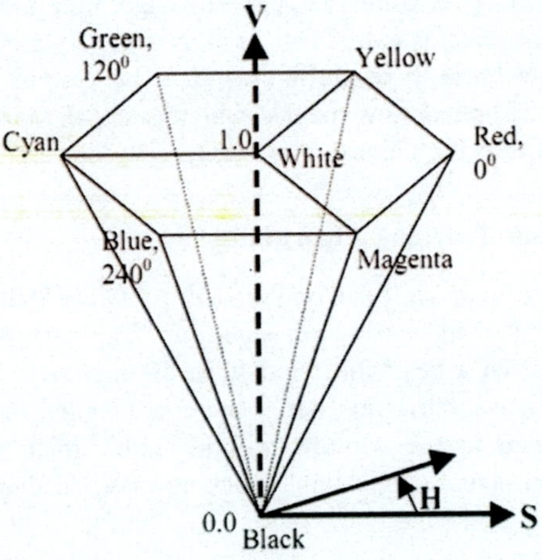

Fig. 2. HSV color model

Unlike the RGB model, the HSV model is an inverted cone model. The model is described in terms of color, depth, lightness and darkness. H stands for color, S stands for depth, and V stands for light and dark. It can be seen as a result of erecting and flattening the central axis of the RGB model as shown in Fig. 2:

Since the three components of the HSV model are more independent than the RGB model, and the RGB model is easily affected by light, the three components of the HSV model are more in line with human visual perception, so many algorithms currently use the HSV model to obtain better stability. In the cam-shift tracking algorithm of this article, the H component is used as the detection feature of the tracking target, which can eliminate the influence of illumination changes.

The basic steps of the cam-shift tracking algorithm are:

(1) Select the current frame image in a series of video frame images, transform its color model from RGB model to HSV model, and extract the color component H;
(2) Select a detection window, establish the color histogram of the H channel of the detection window, and calculate the color probability distribution diagram of the detection window, that is, the back projection diagram;
(3) Select an initial window slightly larger than the detection window as the initial tracking window of the mean-shift algorithm;
(4) Use mean-shift to calculate the position of the centroid of the tracking window, and move the center of the tracking window to the position of the centroid;
(5) Determine whether the center and the center of mass converge, that is, whether the preset accuracy threshold is met. If satisfied, proceed to step (6), otherwise return to step (4);
(6) Perform the tracking of the image target in the next frame, and use the final window size and centroid position obtained in the previous frame as the initial window size and center position of the next frame, and then restart step (3).

The following cam-shift tracking is performed on a group of video data, and the results are shown in Fig. 3:

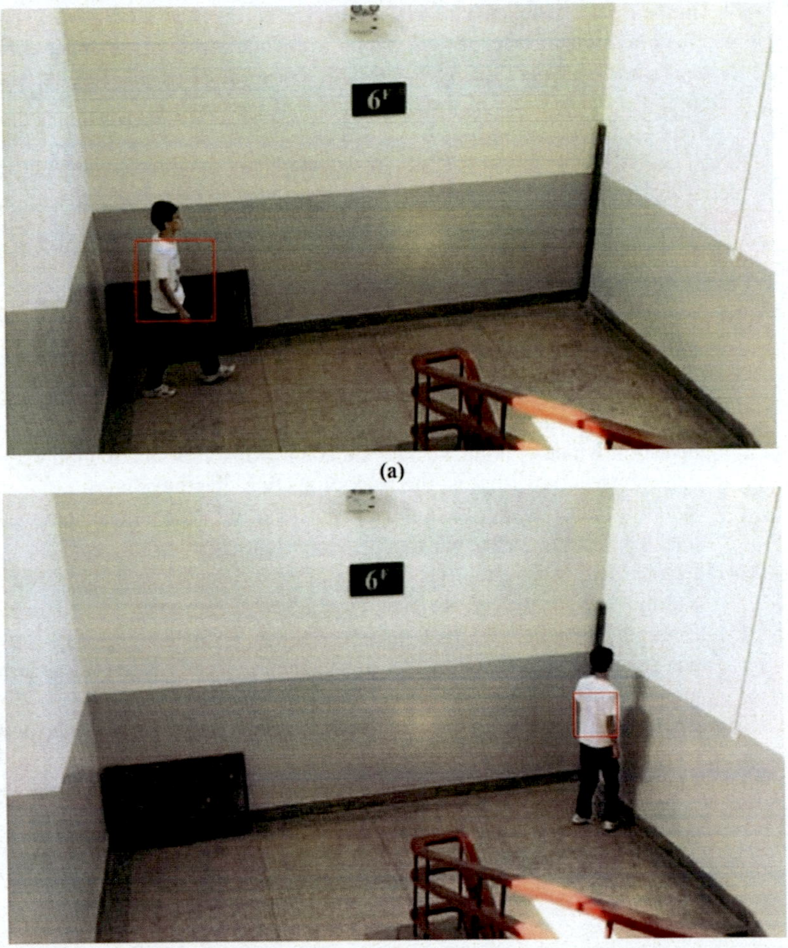

(a)

(b)

Fig. 3. Cam-shift simulation (a). Cam-shift simulation (b)

It can be seen that even if the range selected at the beginning is larger than the moving target, due to the adaptive adjustment of the tracking window of the cam-shift tracking algorithm, the final tracking result can fit the moving target well.

4 Multi-machine Joint Video Tracking Experiment Verification

4.1 Research on Helmet Detection Algorithm

The colors of the helmet are mainly white, red, yellow, and blue, so we can first determine whether there is a helmet in the target area, so as to determine whether the target is wearing a helmet.

However, when the detected target area contains four colors of white, red, yellow, or blue, but not the color of the helmet, but the noise in the background, these noises may be detected as a helmet and cause false alarms or false alarms. Underreporting. In order to solve this problem, this article adds the step of comparing the center of the detected helmet color area with the center of the target area to prevent misjudgment.

Taking the red helmet as an example, the main idea is: first detect the moving target area in the image, and extract the red area. Then calculate the difference Δx and Δy between the center of the moving target area and the center of the red area in the horizontal and vertical directions. Finally, calculate the ratio x' and y' of Δx and Δy to the horizontal width and vertical height of the entire moving target area, and determine whether the helmet is worn normally by comparing with the threshold.

The threshold setting refers to the national standard for adult head size GB/T2428-1998 and the national standard for helmets (GB 2811-89). The vertical direction is set to 0.45–0.48. The smaller the horizontal direction, the better. However, considering the actual situation and the ratio of adult body width to head width, it is set to 0.36.

The basic steps of the helmet detection algorithm are as follows:

1. Detect the moving target in the video image sequence and extract the target area;
2. Calculate the center of the target area and record it as point $i(x, y)$;
3. Detect the red area in the target area and calculate the center of the red area, denoted as the point $r(x, y)$;
4. Calculate the difference Δx and Δy of the point $r\ (x, y)$ and the point $i\ (x, y)$ in the horizontal and vertical directions respectively, and then calculate the ratio of Δx and Δy to the horizontal width and vertical height of the motion area x' and y';
5. Compare x' and y' with the preset threshold range, and classify them as "1" if they are satisfied, otherwise they are classified as "0";
6. Do the AND operation of the above two values to get the final result. If it is 1, it is determined that the helmet is worn normally, and if it is 0, the helmet is worn abnormally, and the alarm can be linked.

4.2 Analysis of Experimental Results

The video data used for the experimental analysis here comes from the actual production video of the simulated substation. According to the experimental results, it can be concluded that the algorithm can basically complete the detection of the wearing state of the workers' helmets in actual production (Fig. 4).

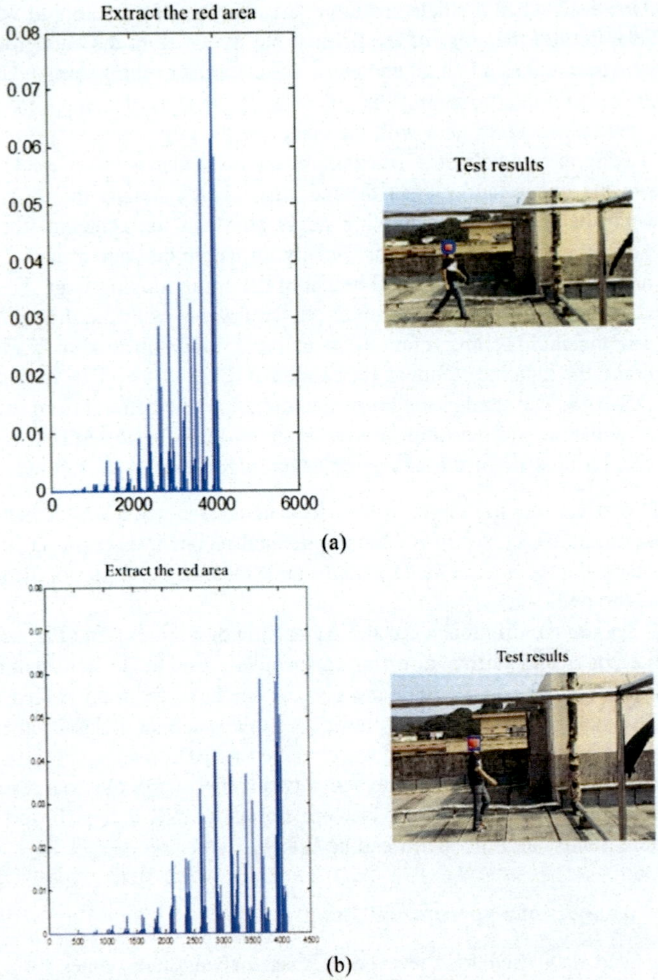

Fig. 4. Safety helmet test results (a). Safety helmet test results (b)

5 Conclusion

This article introduces the basic principles of the mean-shift algorithm and its application in image tracking, and affirms the advantages of mean-shift. However, there are still some shortcomings, which leads to the cam-shift algorithm, and the analysis proves that cam-shift is better than mean-shift. Finally, the method introduced in this article is used in the actual production of safety helmet detection substations. The experimental results show that the method used in this article can basically meet the requirements of substation safety protection.

References

1. Li, H., Zhang, Y., Liang, G.: Application of foreground detection technology in intelligent video monitoring system of substation. In: Proceedings of the 2nd International Conference on Computer Science and Application Engineering, pp. 1–5 (2018)
2. Wang, Q., Wen, B., Wang, X.: Multi-agent based intelligent video monitoring for unattended substation. In: 2008 First International Conference on Intelligent Networks and Intelligent Systems, pp. 515–518. IEEE (2008)
3. Collins, R.T., Lipton, A.J., Kanade, T.: Introduction to the special section on video surveillance. IEEE Trans. Patten Anal. Mach. Intell. **22**(8), 745–746 (2000)
4. Haritaoglu, I., Harword, D., Davis, L.: W4: real-time surveillance of people and their activities. IEEE Trans. Pattern Anal. Mach. Intell. **22**(8), 809–830 (2000)
5. ShuYu, X., Qian, C.: Model and algorithm design of real-time vedio object recognition and counting system. Tsinghua Univ. Pap. **41**(7), 6164 (2001)
6. Liu, S.M., Xu, X.J.: Substation intelligent monitoring system based on pattern recognition. In: Zeng, D. (ed.) Applied Informatics and Communication, vol. 225, pp. 473–478. Springer, Heidelberg (2011). https://doi.org/10.1007/978-3-642-23220-6_60
7. Sreenu, G., Durai, M.A.S.: Intelligent video surveillance: a review through deep learning techniques for crowd analysis. J. Big Data **6**(1), 1–27 (2019)
8. Yang, J., Dang, R., Luo, T., et al.: The development status and trends of unmanned ground vehicle control system. In: 2015 IEEE International Conference on Cyber Technology in Automation, Control, and Intelligent Systems (CYBER), pp. 1946–1952. IEEE (2015)
9. Wren, C.R., Ali, A., Darrell, T., et al.: Pfinder: real-time tracking of the human body. IEEE Trans. Pattern Anal. Mach. Intell. **19**(7), 780–785 (1997)
10. Cui, Y.: Image processing and analysis: mathematical morphology and its applications, vol. 38 (2000)
11. Zhu, Y., Carragher, B., Glaeser, R.M., et al.: Automatic particle selection: results of a comparative study. J. Struct. Biol. **145**(1–2), 3–14 (2004)
12. Kang, W.X., Yang, Q.Q., Liang, R.P.: The comparative research on image segmentation algorithms. In: 2009 First International Workshop on Education Technology and Computer Science, vol. 2, pp. 703–707 IEEE (2009)

Substation Equipment Defect Identification Method Based on Mask R-CNN Algorithm

Hao Sun(✉), Tao Qian, Shuai Zou, Fangwei Zhong, and Bing Han

Xinjiang Information Industry Co., Ltd., Urumqi 830000, Xinjiang, China

Abstract. The structure of power system power network is huge and there are many terminal devices, which is in line with the characteristics of edge computing. It is necessary to use the image recognition algorithm for edge computing to solve the problem of defect recognition of substation equipment. This paper presents a substation equipment defect identification method based on mask r-cnn algorithm. Firstly, the equipment and related defects of substation are introduced. By comparing the defect characteristics and horizontal comparison of algorithms, mask r-cnn is selected as the core algorithm of this paper, and the optimized gfpn network of mask r-cnn is introduced. Finally, the field images are used to test the proposed method to verify its feasibility and accuracy.

Keywords: Mask R-CNN · Substation · Equipment · Defect recognition

1 Instruction

The normal operation of substation equipment is a necessary condition to ensure the continuous and stable operation of the power system. The power system currently used in China includes power generation, transformation, transmission and distribution, and electricity consumption. As an important part of the power grid, the safe and stable operation of the transformation part must be guaranteed [1]. When the substation equipment is connected to the power grid, various equipment defects may occur due to high temperature, heavy overload, lightning strikes, porcelain aging, or changes in the operation mode of the grid. The problem has become a hidden danger of unsafe operation of the power grid. In order to detect equipment defects in substations in time, prevent safety accidents caused by defects in substation equipment, realize efficient management of substation equipment, and improve the efficiency of operation and maintenance of substation equipment, it is reasonable to adopt a machine to conduct automatic inspection of substation equipment. New automation technology for monitoring. In addition, due to the fixed location of the substation equipment and the single background, it is very suitable for edge computing-oriented power system convolutional neural network target detection. A new type of inspection and monitoring method for machine self-learning target detection through image sensors such as substation cameras and inspection robots are indispensable [2, 3].

In order to ensure the intensification of personnel and realize intelligent substation equipment inspection and defect tracking, Zhang Canfeng and others used navigation positioning, pattern recognition and other technologies to integrate substation

Z. Hu et al. (Eds.): AIPE 2021, LNDECT 119, pp. 114–127, 2022.
https://doi.org/10.1007/978-3-030-97064-2_12

intelligent inspection robots equipped with infrared cameras to conduct equipment inspections, and complete its application feasibility through case analysis [4]. In 2019, Liu Jing [5] and others completed a substation inspection robot pan-tilt vision system, and constructed a high-precision visual positioning algorithm system by comparing the matching accuracy of SIFT, SURF, and ORB feature descriptors. Liu Ji [6] and others used the Davidson-Cole model to extract the characteristic parameters β, τ and $\Delta\varepsilon$ to judge the frequency domain aging degree of transformer oil-paper insulation. Zhao Yu [7] and others explored the application of edge computing in the substation system, built an EC-SCADA system, and used edge computing to improve the algorithm of the existing substation monitoring system, and obtained a higher response than the traditional monitoring system. The fast EC-SCADA system has further improved the intelligent management and control of substations, and has better guarantee for the safe operation of substation equipment, making the current substation equipment inspection, defect identification and safety management more intelligent [8, 9].

This paper presents a method for identifying defects in substation equipment based on Mask R-CNN algorithm. First introduce the equipment of the substation and its related defects, and select Mask R-CNN as the core algorithm of this article by comparing the characteristics of defects and the horizontal comparison of algorithms, and introduce the optimized GFPN network of Mask R-CNN. Finally, the method proposed in this paper is tested using the images collected on site to verify its feasibility and accuracy.

2 Method of Time-Series Data Curve Alignment

2.1 Defects of Substation Equipment

Long-term operation of substation equipment in a high-voltage environment, whether it is the heat generated by the electrical load generated by daily work or the aging caused by sunlight and weathering in the natural environment, can cause serious defects to the substation equipment. When substation equipment is prone to defects, the safe and stable operation of the power system will be at risk. Therefore, it is necessary to conduct real-time inspection and analysis of the defects of substation equipment to avoid defects generated by substation equipment from affecting the reliability of power supply of the power grid.. The following methods are only applicable to the defect identification and classification of this article to classify the defects of the existing substation equipment.

(1) **Visible Defects in Appearance**

In the actual situation of daily production, the substation usually carries a large amount of load. Due to the heat generated inside the load and the influence of external forces such as rainfall, wind, sun, and animals in the natural environment, the multi-substation equipment is easy to change. Appearance defects visible to the naked eye.

1) *Leaking oil*

The function of insulating oil for substation equipment is to insulate and dissipate heat, and there is usually an oil mark on the oil-using equipment. Due to the aging of the rubber, the loss of elasticity and the uneven internal crimping or the small part of the explosion will cause the oil standard to leak. Due to the limitation of welding technology, the body of the transformer and the radiator are also prone to oil leakage.

2) *Porcelain bottle burst*

Due to the unclean source of the ambient air in which the substation equipment works, the wind source contains a lot of moisture, impurities, etc. Flashover discharge occurs during the switching operation of the substation equipment, and the air containing moisture will also induce ionization under high pressure and cause tension. Arc, the moving and static contacts are damaged, and finally the porcelain bottle bursts.

3) *Isolating switch is not in place*

The isolating switch is also called the knife switch, which is used as a switching device without arc extinguishing function in the production site. There are two ways to drive the opening and closing in the process of daily use, manual and electric. In a highly automated substation, it is prone to problems of inadequate opening and closing due to rusty shafts or insufficient power supply in the secondary circuit. In daily production, the phenomenon of inadequate closing is the majority.

4) *Damaged insulation sheet*

Due to the oxidation problem caused by the normal working environment, the insulation sheet of the substation equipment is prone to breakage after aging. After the insulation sheet is damaged, the ground insulation effect of the equipment will decrease, which poses a great safety hazard [12].

5) *Broken strands of the diversion line*

In the substation, the diversion wire is usually a twisted wire method, and the installation process is prone to over-tight or over-installation due to the installation threshold problem. If the diversion wire is too loose, it is easy to cause a short circuit between phases due to problems such as wind and wire swing. If the diversion wire is too tight, it is easy to cause strand breakage and the circuit cannot operate normally.

(2) **Fever Defect**

In the daily operation of substation equipment, many defects exist in the interior, and it is impossible to visually tell from the appearance whether the substation equipment in operation is defective. Due to the thermal effect of current, most of the substation equipment works under high load, and part of the electrical energy is converted into thermal energy for external dissemination. The defective part of the substation equipment will generate a large current, and its heat generation is much higher than the normal operating state, and the heating defect can be detected by an infrared thermometer. Usually, the heating defects are mostly found in the wiring, and the isolation switch is the most common. The main causes are poor contact, rusty equipment, long-term heavy-overload operation, etc. [11].

2.2 Substation Defect Recognition Algorithm

There are many types of defects in substation equipment, and each type of substation equipment may have different types of defects. However, the location of the substation equipment is fixed and the background is single, which is relatively easy to train in the visual neural network of the power system. In order to better monitor the related defects of substation equipment, the algorithm for the recognition of substation equipment defects should have a high-depth model and high-precision recognition accuracy. Image recognition based on different environmental backgrounds, different times, and different angles under the status of substation equipment should also achieve better accuracy. Therefore, the algorithm for defect recognition of substation equipment should meet the following conditions:

(1) **High Depth Model**

The algorithm requires high-precision and effective recognition of the normal state and defect state of the substation equipment. The defects in the substation equipment are not fixed and there are many types. It is necessary to complete the feature extraction and image classification of the target equipment under the interference of other equipment in the background, Semantic segmentation and other work. Therefore, the algorithm applied to the defect recognition of substation equipment should have a deeper convolutional neural network, so as to complete the defect detection that requires higher accuracy.

(2) **Precise Semantic Segmentation and Feature Fusion Capabilities**

Under normal circumstances, the scope of defects of substation equipment is not large, and there are a lot of small defects. In this case, the defect recognition algorithm for substation equipment requires precise semantic segmentation capabilities. Therefore, the defect recognition algorithm of substation equipment must not only have accurate semantic segmentation ability but also feature fusion ability. The algorithm needs to merge the simple features of the lower level with the semantic features of the high level to reduce the loss of defect information.

(3) **Screen Irrelevant Candidate Domains**

For the image recognition process of convolutional neural network, the algorithm will form a proposed region (Region Proposed) for the input image, but in general, most of the proposed regions generated are not consistent with the target detection part, which belongs to an invalid suggestion domain. The invalid suggestion domain not only has a certain impact on the speed of the algorithm, but may also reduce the overall recognition accuracy of the algorithm. Almost all single-stage algorithms do not add filtering to the suggested domains, and the two-stage algorithm starts to filter the suggested domains.

For the substation equipment, its location is fixed and the background is unchanged. The equipment that collects image data in the substation is mostly a fixed camera.

Therefore, you can choose the Mask R-CNN algorithm with strong semantic segmentation ability and outstanding accuracy. For the information of the substation equipment of the picture data, the Mask R-CNN algorithm with a recognition rate of only 5FPS can also handle it well. The following data set training and model sampling for defect recognition of substation equipment are implemented based on the Mask R-CNN algorithm.

3 Mask R-CNN Algorithm and Principle

3.1 Mask R-CNN Algorithm

The Mask R-CNN algorithm outputs a Binary Mask value for each suggestion domain, and the classification ability in the Mask R-CNN algorithm is based on Mask Prediction. The Mask R-CNN algorithm is different from the Fast R-CNN algorithm and the Faster R-CNN algorithm in its RoI Align method. Other algorithms use the RoI Pooling operation, while the RoI Align method in the Mask R-CNN algorithm It involves bilinear interpolation. The block diagram of Mask R-CNN can be shown in Fig. 1.

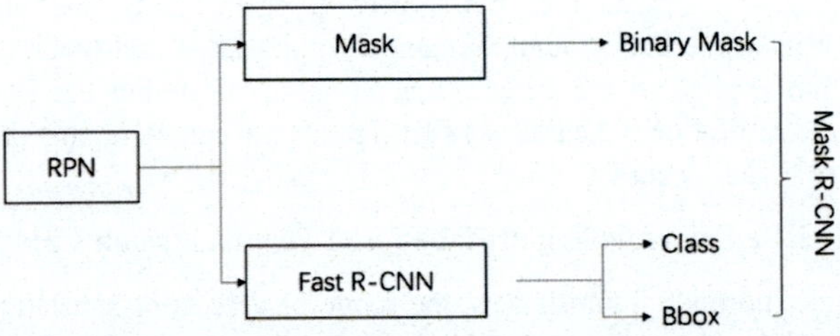

Fig. 1. Simple schematic diagram of Mask R-CNN

(1) **RoI Align Method**

1) *Bilinear interpolation*

Interpolation problems are common in numerical analysis and other fields, and the commonly used linear interpolation formulas are shown in Eq. (1).

$$y = \frac{x - x_0}{x_1 - x_0} y_1 + \frac{x_1 - x}{x_1 - x_0} y_0 \tag{1}$$

The bilinear interpolation problem is more complicated than the ordinary linear difference problem. The bilinear interpolation problem can be expressed as Eq. (2–4).

$$f(R_1) \approx \frac{x_2 - x}{x_2 - x_1} f(Q_{11}) + \frac{x - x_1}{x_2 - x_1} f(Q_{21}) \tag{2}$$

$$f(R_2) \approx \frac{x_2 - x}{x_2 - x_1} f(Q_{12}) + \frac{x - x_1}{x_2 - x_1} f(Q_{22}) \tag{3}$$

$$f(P) \approx \frac{y_2 - y}{y_2 - y_1} f(R_1) + \frac{y - y_1}{y_2 - y_1} f(R_2) \tag{4}$$

The thinking of bilinear interpolation is used to solve the problem that the RoI area in the Mask R-CNN algorithm cannot be pixel-aligned with the original image.

2) *Suggested domain matching alignment*

RoI Pooling operations often have non-integer area coordinates. Both the x and y values in the position function are decimals. RoI Align is an optimization method that can solve this problem. RoI Align divides the anchor points in the proposed domain equally, and calculates the bilinear interpolation result from each sampling point to its nearest grid point.

(2) **Loss Function**

The loss function that will appear during the operation of the convolutional neural network is mentioned above. The loss function is used as an intuitive embodiment of the training results of the convolutional neural network. The lowest point of the function represents the optimization of the model. The loss function of the Mask R-CNN algorithm is shown in Eq. (5).

$$L = L_{cls} + L_{box} + L_{mask} \tag{5}$$

In the formula, L_{cls} represents the classification error, L_{box} represents the bounding box error, L_{mask} represents the segmentation error, and L represents the loss function of the Mask R-CNN algorithm.

(3) **Algorithm Application**

The Mask R-CNN algorithm is proposed for instance segmentation (Instance Segregation). Compared with semantic segmentation, the accuracy of instance segmentation is more accurate. Instance segmentation not only divides each pixel of the detected target, but also divides the pixels of different targets. The Mask R-CNN algorithm also has a layer of Mask branch on the basis of classifying the target and anchoring the frame to segment the specific contour of the target. In the actual application of engineering, you can choose to change the classification of Mask R-CNN to the actual required classification to complete the detection of the target in the image. At present, Mask R-CNN has even been applied to the target recognition of the 6D module to identify the three-dimensionality of the target [13, 14].

3.2 Mask R-CNN for Edge Computing

In the process of Mask R-CNN algorithm for the defect detection of substation equipment, in order to better compress the model, the convolution kernel can be simplified by first performing 1 × 1 convolution on the feature layer generated by the convolution to simplify the feature map. Then perform the corresponding convolution operation to return the feature value.

3.2.1 Correlation Coefficient Derivation

In the Mask R-CNN algorithm process based on the GFPN framework, the model compression part should be mainly aimed at its CNN part. For the simplification in the convolutional neural network, it should be tailored according to the network characteristics, and the weight ratio should be selected as redundant or relative. Cut the unimportant parts. The main idea of tailoring the entire CNN network is shown in Fig. 2.

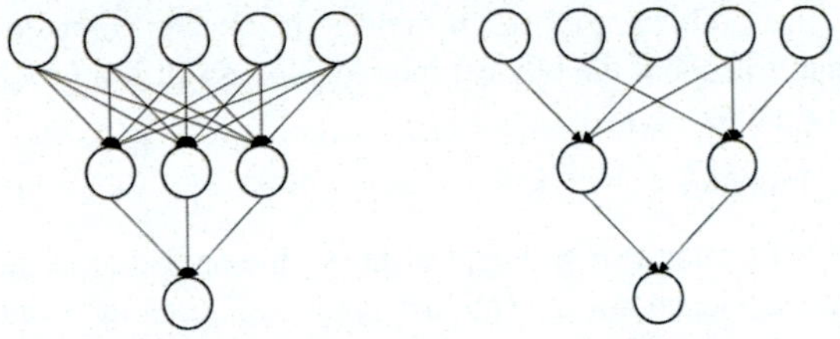

Fig. 2. CNN network weight pruning diagram

As for weight pruning, it is not completely random to trim. The influence of each weight on the algorithm's ability to recognize defects in substation equipment should be calculated to determine whether it should be trimmed. In order to avoid the loss of algorithm network accuracy due to weight pruning, the pruning process should satisfy the formula (6).

$$\min(\Delta) = \min_{W^{\rho}} |E(D|f, W^{\rho}) - E(D|f, W)| \tag{6}$$

Where W is the weight of the fully connected layer, D is the deep learning training set, E is the loss function, W^{ρ} is the weight after pruning, and Δ is the loss function distance.

Under the condition of ensuring accuracy and efficiency, the weight neurons that should be tailored are screened by formula (3.9) to obtain a more simplified model and faster calculation speed. Reduce the redundancy of the fully connected layer while ensuring the Mask R-CNN algorithm's ability in target detection.

3.2.2 Weight Parameter Sharing

For the fully connected layer, the sharing of experimental weight parameters can also effectively compress the algorithm model accordingly. The fully connected layer is divided into two cases: the number of neurons in the input layer and the output layer is equal and not equal. For the case where the number of neurons is equal, researchers usually construct a circulant matrix by phase shifting the weights of the connections between neurons, and finally construct a circulant matrix of weights. The circulant matrix of weights is the weight parameter. An important way of sharing. However, the number of neurons in the output layer of the fully connected layer in the Mask R-CNN algorithm is usually inconsistent with the number of neurons in the input layer, so the toeplitz matrix needs to be used to plan the weights for the sharing of weight parameters. The planning method of the toeplitz matrix is to take the same value from the upper left to the lower right of the weight value, and the adjacent rows recursively complete the matrix planning.

In the process of weight parameter sharing, not only the parameters of the fully connected layer can be shared, but the convolution kernel operation of the convolutional neural network in the convolution process can also be weight shared. For a convolution kernel with dimensions m × n × h × w (where m is the number of input feature layers, n is the number of output layers, h is the height of the convolution kernel, and w is the width of the convolution kernel), from The input part cyclically shifts the convolution kernel along the w direction, and its compression factor satisfies formula (7).

$$M = \frac{m \times n \times h \times w + m}{m \times h \times w + m} \tag{7}$$

Where m is the number of output layer images, n is the number of input layer images, h is the height of the convolution kernel, and w is the width of the convolution kernel. For this method of convolution kernel cyclic shift to reduce the amount of network parameters, it can be seen that the compression factor is large, and the decrease in network accuracy has no major impact.

4 Mask R-CNN Substation Defect Recognition Experiment Verification

4.1 Recognition Result of Substation Equipment Set

Take the image collection sample of substation equipment as an example, as shown in Fig. 3.

Fig. 3. Sample set of substation equipment

The Mask R-CNN network is used to conduct centralized training on the image set of substation equipment, and the recognition result corresponding to the sample set shown in Fig. 3 is taken as an example. The recognition result of the Mask R-CNN network using the GFPN framework is shown in Fig. 4.

From the comparison of Fig. 3 and Fig. 4, it can be seen that the Mask R-CNN algorithm can not only identify and classify substation equipment targets, but also complete the instance segmentation content of substation equipment. Mask R-CNN carries the relevant identification of substation equipment with its outstanding instance segmentation ability.

Fig. 4. Example of identification results of substation equipment

In order to test the recognition of defects in substation equipment based on the Mask R-CNN algorithm, in Sect. 3.2, the Mask R-CNN algorithm is applied to the new sample set generated on the defect data set of substation equipment through the GAN network for in-depth training and test.

4.2 Defect Recognition Results

Since there are not many cases of defects in substation equipment, there is a certain lack of training data even after the stored defect sample set of substation equipment is generated and increased by the GAN network. Existing defect samples of substation equipment are shown in Fig. 5. This article selects the more typical images in the sample set as an example. The defect of the substation equipment in Fig. 5 is the oil leakage fault of the transformer.

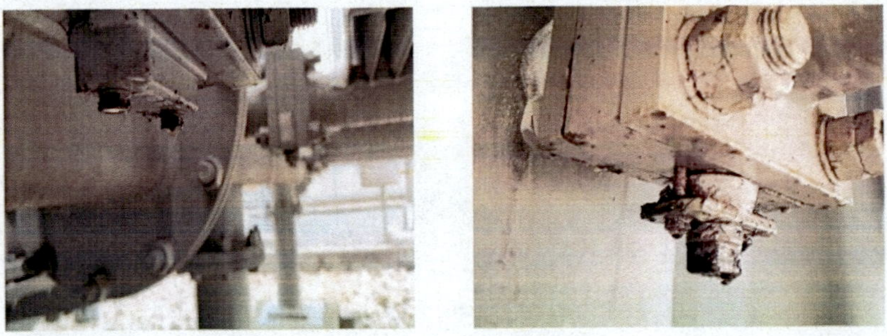

Fig. 5. Defect sample

Using the Mask R-CNN network for defect recognition and detection, the output result of the Mask R-CNN algorithm is shown in Fig. 6. Due to the Bounding Box area division for the defects of the substation equipment under the Mask R-CNN algorithm in the calibration process, the output instance segmentation content only contains a small area of the equipment.

The confidence levels of defect recognition in Fig. 6 are 0.530 and 0.712, which are insufficient for image recognition of defects in substation equipment. It can be seen from the Softmax probability output of substation defect recognition in Fig. 10 that because of the small number of sample sets and insufficient training, the Mask R-CNN algorithm of the GPFN architecture has insufficient confidence in the recognition of substation equipment defects, and its accuracy is still insufficient. Need to improve, for the results need to continue to improve.

Fig. 6. Example of defect recognition

4.3 Results Irovement

According to the result of the defect recognition of substation equipment based on the Mask R-CNN algorithm, in order to improve the detection accuracy of the algorithm, the defect of substation equipment can be judged more accurately. The identification process is divided into two steps-the first step is to use the substation equipment recognition model to perform image recognition on the substation equipment under daily operating conditions, and use the Softmax probability output result as the two-class criterion to preliminarily judge whether it has a fault. The second step is to choose to put the faulty image into the defect sample data set for model deepening training, so as to achieve better defect recognition ability. When the defect criterion is performed in the first step, the classification situation is shown in Table 1.

Table 1. Algorithm recognition classification

	Algorithm identified as defect	Algorithm recognized as normal
Real defect	Normal defect	Error defect
Real operation	Normal operation	Error operation

For the defect criterion of substation equipment, this paper introduces the definition threshold ρ and defect criterion δ. When δ is less than ρ, it is determined that the substation equipment is not defective and belongs to the normal operating state. When δ is greater than ρ, it is judged to be defective and belongs to a fault state, and the fault image is put into the defect sample data set of substation equipment for model training. The calculation expression is shown in formula (8).

$$\delta = |P_0 - P_s| \tag{8}$$

In the formula, δ represents the defect criterion value, P_0 represents the Softmax output value of the image being inspected, and P_s represents the stable Softmax output value of the similar equipment in the sample set.

The criterion is carried out according to formula (8), and linear interpolation, spline interpolation, quadratic interpolation, and quartic function fitting are performed for the values of different judgment thresholds ρ, as shown in Fig. 7.

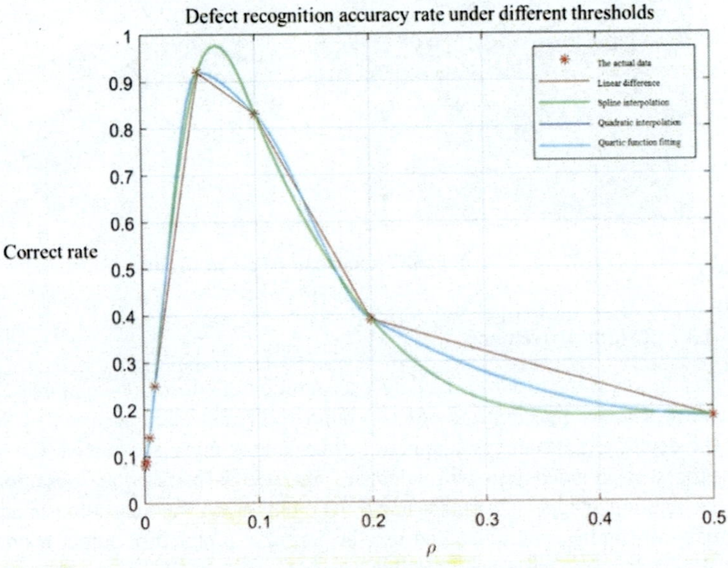

Fig. 7. Function fitting

The thresholds for determining when the function reaches the highest point in the figure are 0.0505, 0.0657, 0.0505, and 0.0505, respectively. The average value of the judgment threshold is 0.0543. In this paper, {TP, FP} in the real defect is recorded as the algorithm sensitivity, and {TN, FN} in the real operation is recorded as the algorithm specificity, and the judgment threshold ρ is 0.0543. Use this as the characteristic-sensitive curve diagram of the algorithm, and put the artificial image classification and average situation into the reference, as shown in Fig. 8.

It can be seen from Fig. 8 that the accuracy of the defect criterion method with a judgment threshold of 0.0543 is higher than the average accuracy of artificial image classification, which verifies the feasibility of the improved method. Judging from the 25 image classification results of 1117 defective images of substation equipment completed manually, the average value obtained is compared with that of the Mask R-CNN algorithm under the GFPN framework for defect recognition in the defect data set of substation equipment. Its recognition accuracy has the ability to be put into production. The identification of defect data of substation equipment will also be gradually improved as the number of defective equipment is determined, which has practical engineering value.

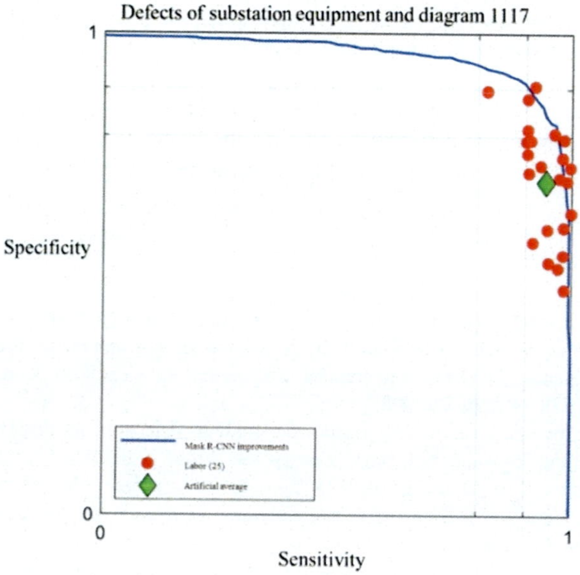

Fig. 8. Comparison of improved algorithm and manual accuracy

5 Conclusion

As an important node of the power system, the substation plays a vital role in the stable operation of the power system. The hidden dangers of defects in substation equipment should be eliminated. Substation operation and maintenance personnel cannot monitor in real time during daily inspections. Therefore, image sensors based on the perception layer of the ubiquitous power Internet of Things should meet the ability to recognize defects in substation equipment. The emergence of edge computing is to solve the problem that a large amount of image data cannot be fully processed by a single device. The power system has a huge power network architecture and most of its terminal devices are in line with the characteristics of edge computing. Therefore, an image recognition algorithm oriented to edge computing is needed to solve the defect recognition problem of substation equipment.

Since the defects of substation equipment are usually irregular and uncertain defects, the Mask R-CNN algorithm, an image recognition algorithm with instance segmentation capability, is chosen to carry the task of identifying defects in substation equipment for edge computing. Since the location and image background of the substation equipment are fixed, the shortcomings of the insufficient operation speed of Mask R-CNN can be overcome. This paper verifies the feasibility of the improved Mask R-CNN algorithm and proposes a basis for determining defects based on the two-class classification of Softmax output results. However, at present, for the relatively hidden and rare defect identification of substation equipment, the algorithm is not well-fitted and the confidence is not high.

The accuracy and speed of the algorithm for identifying sudden, rare, and highly concealed defects in substation equipment still needs to be improved. This article is limited by the sample set, and the accuracy of the defect judgment basis and the fitting process are still worth studying. In engineering practice, we should keep hoarding typical data sets of defective images of substation equipment and continue to train the Mask R-CNN algorithm of the FPN framework to improve the accuracy even more.

References

1. Kinsht, N.V., Petrun'ko, N.N.: The experience of inspection of a technical condition of the HV equipment on substation by a method of registration of the own electromagnetic radiations. In: 2008 International Conference on Condition Monitoring and Diagnosis, pp. 738-740. IEEE (2008)
2. Guelpa, E., Verda, V.: Automatic fouling detection in district heating substations: Methodology and tests. Appl. Energy **258**, 114059 (2020)
3. Mishra, D.K., Dhara, S., Koley, C., et al.: Self-organizing feature map based unsupervised technique for detection of partial discharge sources inside electrical substations. Measurement **147**, 106818 (2019)
4. Hajian-Hoseinabadi, H.: Reliability and component importance analysis of substation automation systems. Int. J. Electr. Power Energy Syst. **49**, 455–463 (2013)
5. Liu, X., Xu, K., Zhou, P., et al.: Surface defect identification of aluminium strips with non-subsampled shearlet transform. Opt. Lasers Eng. **127**, 105986 (2020)
6. Kristan, M., et al.: The visual object tracking VOT2014 challenge results. In: Agapito, L., Bronstein, M., Rother, C. (eds.) ECCV 2014. LNCS, vol. 8926, pp. 191–217. Springer, Cham (2014). https://doi.org/10.1007/978-3-319-16181-5_14
7. Yuan, Z.W., Zhang, J.: Feature extraction and image retrieval based on AlexNet. In: Eighth International Conference on Digital Image Processing (ICDIP 2016), vol. 10033, p. 100330E. International Society for Optics and Photonics (2016)
8. Song, Y.D., Lewis, F.L., Polycarpou, M., et al.: Guest editorial special issue on new developments in neural network structures for signal processing, autonomous decision, and adaptive control. IEEE Trans. Neural Netw. Learn. Syst. **28**(3), 494–499 (2017)
9. Tyagi, K., Manry, M.: Multi-step training of a generalized linear classifier. Neural Process. Lett. **50**(2), 1341–1360 (2019)
10. Steinhurst, B., Teplyaev, A.: Spectral analysis on Barlow and Evans' projective limit fractals. J. Spectr. Theory **11**(1), 91–124 (2021)

11. Yang, Z., Yabansu, Y.C., Al-Bahrani, R., et al.: Deep learning approaches for mining structure-property linkages in high contrast composites from simulation datasets. Comput. Mater. Sci. **151**, 278–287 (2018)
12. Mahendran, A., Vedaldi, A.: Understanding deep image representations by inverting them. In: Proceedings of the IEEE Conference on Computer Vision and Pattern Recognition, pp. 5188–5196 (2015)
13. Lin, T.Y., Maire, M., Belongie, S., et al.: Microsoft coco: common objects in context. In: Fleet, D., Pajdla, T., Schiele, B., Tuytelaars, T. (eds.) ECCV 2014. LNCS, vol. 8693, pp. 740–755. Springer, Cham (2014). https://doi.org/10.1007/978-3-319-10602-1_48
14. Pineau, E., Lelarge, M.: InfoCatVAE: representation learning with categorical variational autoencoders. arXiv preprint arXiv:1806.08240 (2018)

Comprehensive Evaluation Method and Index System for Electric-Hydrogen-Storage Integrated Energy Network

Shuai Wang[1], Chaofan Zhao[2], Haijun Liu[3], Runhao Gao[1], Siwei Liu[1], and Xiaoling Su[2](✉)

[1] Zhangjiakou Chongli District Power Supply Company, State Grid Jibei Electric Power Co., Ltd., Zhangjiakou 075000, China
[2] Qinghai Key Lab of Efficient Utilization of Clean Energy, Qinghai University, Xining 810016, Qinghai, China
suxiaoling@qhu.edu.cn
[3] State Key Laboratory of Advanced Power Transmission Technology, Global Energy Interconnection Research Institute Co., Ltd., Beijing, China

Abstract. Integrated energy network (IEN) plays an important role in energy efficiency and low carbon emissions development. In order to host the low-carbon 2022 Winter Olympic Games in Beijing this paper proposes comprehensive evaluation method and index system for electric-hydrogen-storage integrated energy network in Chongli Winter Olympics zone based on traditional evaluation methods to evaluate this integrated energy network which is a double-level, 6-indicator evaluation system and the corresponding weight of each index are calculated automatically. Fuzzy synthetic evaluation model is developed considering the uncertainties of renewable energy sources (RESs) to improves accuracy of this evaluation set model. The simulation results prove that the proposed multi-dimensional evaluation model forms a scientific and reliable evaluation system under uncertain operating conditions for electric-hydrogen-storage integrated energy network. The benefits of integrated energy network are evaluated from six perspectives, such as economy and low carbon, to provide theoretical basis for future planning and evaluation of integrated energy network.

Keywords: Integrated energy network · Fuzzy synthetic evaluation model · Scenario reduction · Multi-dimensional evaluation · Hydrogen energy

1 Introduction

China proposed its 14th Five-Year Plan and the Vision Target 2035 on October 29th 2020 to accelerate the green and low-carbon development strategy. Carbon emission reduction plans have been formulated according to this plan, in order to decrease the intensity of carbon emissions and reach peak carbon emissions by 2030. Under this background of implementing the goal of carbon neutralization, ensuring energy security, and accelerating energy transformation, renewable energy installation capacity in China is increasing rapidly.

In order to host the low-carbon 2022 Winter Olympic Games in Beijing, low-carbon industries such as sports, culture, tourism and leisure have developed significantly, which

Z. Hu et al. (Eds.): AIPE 2021, LNDECT 119, pp. 128–137, 2022.
https://doi.org/10.1007/978-3-030-97064-2_13

also has optimized the ecological environment. All these efforts will promote and upgrade domestic industrial transfer, and drive the domestic green economic growth in China. There are steps such as build the world's first large-scale photovoltaic and wind power station with four-end flexible DC network, adopt carbon dioxide across critical direct refrigeration ice first time have been taken for natural ecological protection, form green venues and the global 2022 Beijing Winter Olympics "green, sharing, open, clean" promise.

Reference [1] has innovated the model structure of the energy hub, which comprehensively considers the electric, thermal and cold storage and discharge, coupling a variety of energy, and realizes energy collection, saving and distribution. Based on the three subnetwork systems of electricity, gas and heat, Reference 1 establish a new energy absorption potential analysis model, comprehensively considering the coordinated operation of heat storage, electric boilers and gas-fired boilers. Reference [3] choose the efficiency evaluation index from the perspective of technical and economic evaluation, and explains the meaning of each evaluation index, calculation formula, application scenarios and restrictions, etc., established the foundation of complete investment benefits, but the system index mostly focused on the evaluation of power energy, less gas, heat, cold or new energy indicators, ignore the energy diversity. Reference [4] analyzes the benefits of comprehensive energy system from the perspective of environmental pollution, thermal performance and economic benefits, puts forward the calculation method of operating period thermal economy of natural gas distributed energy stations, and comprehensively calculates and evaluates the thermal economy indicators of different distributed energy unit schemes. Reference [5] evaluates the economic, energy and environmental benefits of the comprehensive energy system, namely the "3E" evaluation index system, and uses the entropy method to weighted the total benefit score of the comprehensive energy system. Reference [6] independently analyzed the three indicators of energy, economy and environment and concluded that compared with traditional energy systems, integrated energy systems have higher energy and environmental benefits, while the economic benefits need to be strengthened. However, most papers [7–10] mainly analyze the system benefits from terms of economic benefits, and do not consider the changes in operating benefits caused by low carbon benefits and uncertain new energy output.

Therefore, this paper proposes comprehensive evaluation method and index system for electric-hydrogen-storage IEN in Chongli Winter Olympics zone based on traditional evaluation methods to evaluate this IEN which is a double-level, 6-indicator evaluation system and the corresponding weight of each index are calculated automatically. These two primary indexes of low carbon efficiency and operational efficiency are added to the original evaluation system based on the conventional evaluation system [11, 12] and the renewable energy characteristic [13, 14] in Chongli using the analytic hierarchy process and geometric weighted averaging method.

2 Construction of IEN Evaluation Index System

Combined with the characteristics of electricity-hydrogen-storage IEN, this paper divides the evaluation system into low-carbon benefits, economic benefits, technical benefits, operation benefits, environmental benefits, social benefits, and further subdivided in these six aspects, divided into 20 secondary indicators, to build a complete system comprehensive benefit evaluation system.

Taking the weight of comprehensive benefits as an example, using the analytic hierarchy process and the geometric weighted averaging method, multiple experts are asked to score the index, the comprehensive benefits judgment matrix is shown in the Table 1:

Table 1. Comprehensive benefit judgment matrix and weight calculation results

Comprehensive benefit	Low carbon benefits	Economic benefit	Technical benefits	Operating benefits	Environmental benefits	Social benefits	W_i
Low carbon benefits	1.00	1.00	2.00	2.00	2.00	3.00	0.2563
Economic benefit	1.00	1.00	2.00	2.00	3.00	3.00	0.2714
Technical benefits	0.50	0.50	1.00	1.00	2.00	2.00	0.1502
Operating benefits	0.50	0.50	1.00	1.00	2.00	2.00	0.1502
Environmental benefits	0.50	0.33	0.50	0.50	1.00	1.00	0.0896
Social benefits	0.33	0.33	0.50	0.50	1.00	1.00	0.0822

Table 2. Weight table of comprehensive benefit evaluation index of electricity-hydrogen-storage IEN

Code layer	W_i	Index layer	W_{ij}	Code layer	W_i	Index layer	W_{ij}
Technical benefits	0.1502	System operation efficiency	0.4431	Operating benefits	0.1502	Service life of equipment	0.2411
		System operation and maintenance level	0.3875			Occupy benefits	0.2101
		System technical reliability	0.1694			Comprehensive network loss rate	0.5488
Economic benefits	0.1502	Initial investment	0.2280	Low carbon benefits	0.2714	Carbon cost	0.4
		Annualized cost	0.1847			Carbon emission reduction	0.6
		Levelized cost of energy	0.1966	Environmental benefits	0.0896	Pollutant emissions	0.4
		Net present value	0.1201			Clean energy substitution rate	0.6
		Internal rate of return	0.0857	Social benefits	0.0822	Slow construction ability	0.2515
		Dynamic investment pay-back period	0.0648			Promote the degree of system development	0.5897
		Profit ratio	0.1201			Number of jobs invested by the unit	0.1589

Congruency index of the judgment matrix $CI = 0.0090 < 1.24$, This matrix has consistency [15].

The index weight calculation of low carbon, technical, operational benefits, environmental protection and social benefits is the same as that of comprehensive benefits. The final weight results are shown in the Table 2.

3 Consider the Random Perturbations and the Applicability Analysis

The development of new energy is playing a more and more important role in ensuring power supply and energy transformation. However, new energy output fluctuations will lead to the system node voltage, line current uncertainty, thus affecting the accuracy of the IEN evaluation. This paper adopts the reverse scenario reduction method to obtain by random sampling the fluctuation scene of wind power and photovoltaic power, and the integrated energy system fault state scene through the random sampling of equally distributed Monte Carlo simulation method.

The actual output of wind power and photovoltaic power corresponding to each typical scenario is expressed as follows:

$$P_m^{wd} = \begin{cases} p_1^{wd} & 0 \leq R_m^{wd} \leq Prob_1^{wd} \\ p_s^{wd} & \sum_{i=1}^{s-1} Prob_i^{wd} < R_m^{wd} \leq \sum_{i=1}^{s} Prob_i^{wd} \end{cases} \tag{1}$$

where, R_m^{wd} is random number within the [0, 1] interval; $Prob_i^{wd}$ is The probability of Typical scene i; p_s^{wd} is expected output under scene s12.

Considering that the device fault is a small probability event, this paper adopts Monte Carlo simulation method based on equal dispersion sampling to solve the generation problem of the equipment fault space, divide [0, 1] interval into k subspace, For the k-th fault subspace, the state of device $I_{k,i}$ can be expressed by formula (2):

$$I_{k,i} = \begin{cases} 0 & other \\ 1 & \frac{(k-1)}{K} \leq R_i^{fail} \leq \frac{(k-1)}{K} + Prob_i^{fail} \end{cases} \tag{2}$$

And so on, the status space for the entire system equipment fails can be expressed as $I_k = [I_{k,1}, I_{k,2}, \ldots, I_{k,N}]$.

Considering the system failure caused by the new energy output fluctuations and the resonance of the power electronic equipment, the evaluation system in the above chapter is supplemented, which greatly improves the accuracy of this evaluation system in the uncertain working conditions.

4 Multi-dimensional Evaluation Model of IEN

The fuzzy comprehensive evaluation method is selected to conduct the comprehensive benefit evaluation of the electricity-hydrogen-storage IEN. After establishing the evaluation index system and determining the weight of the evaluation index, the evaluation set also needs to be constructed, as shown in formula (3):

$$\mathrm{V} = \{v_1, v_2, v_3, v_4, v_5\} = \{\mathrm{A}, \mathrm{B}, \mathrm{C}, \mathrm{D}, \mathrm{E}\} \tag{3}$$

The evaluation matrix R was constructed based on degree of membership function matching the evaluation objective, as shown in formula (4):

$$R = \begin{bmatrix} R_1 \\ R_2 \\ \vdots \\ R_m \end{bmatrix} = \begin{bmatrix} r_{11} & r_{12} & \cdots & r_{1n} \\ r_{21} & r_{22} & \cdots & r_{2n} \\ \vdots & \vdots & \ddots & \vdots \\ r_{m1} & r_{m2} & \cdots & r_{mn} \end{bmatrix} \tag{4}$$

The evaluation matrix R fuzzy transforms the weight of each index to form the fuzzy comprehensive evaluation model B, as shown in formula (5):

$$B = R \cdot W = \begin{bmatrix} r_{11} & r_{12} & \cdots & r_{1n} \\ r_{21} & r_{22} & \cdots & r_{2n} \\ \vdots & \vdots & \ddots & \vdots \\ r_{m1} & r_{m2} & \cdots & r_{mn} \end{bmatrix} \cdot (w_{11}, w_{12}, \cdots, w_{1n}) = (b_1, b_2, \cdots, b_m) \tag{5}$$

When determining degree of membership function, the main method is the assignment method, that is, setting the corresponding coefficient according to the experience, and the complexity of the power system data will make the result have great error. According to the above analysis, the method of standardizing the data is proposed.

The standardization of the attribute value should first classify the attribute value, and this article divides it into benefit type, cost type, certainty type and interval type:

1) Standardization method for certainty type values:

$$r_{ij} = \begin{cases} \dfrac{\max a_{ij} - a_{ij}}{\max a_{ij} - \min a_{ij}} & a_{ij} \text{ is Benifit index} \\ \dfrac{a_{ij} - \min a_{ij}}{\max a_{ij} - \min a_{ij}} & a_{ij} \text{ is Cost index} \end{cases} \tag{6}$$

where, a_{ij} is the attribute value, r_{ij} is normalized values.

2) Standardization method for interval type values

$$r_{ij} = \begin{cases} [\frac{a_{ij}^-}{\max a_{ij}^+}, \frac{a_{ij}^+}{\max a_{ij}^+}] & a_{ij} \text{ is Benifit index} \\ [1 - \frac{a_{ij}^+}{\max a_{ij}^+}, 1 - \frac{a_{ij}^-}{\max a_{ij}^-}] & a_{ij} \text{ is Cost index} \end{cases} \tag{7}$$

where, $a_{ij} = [a_{ij}^-, a_{ij}^+]$ is attribute value, $r_{ij} = [r_{ij}^-, r_{ij}^+]$ is normalized values.

If the economic benefit analysis is obtained as B_1 = (0.008, 0.105, 0.424, 0.380, 0.084) in other words, in the B_1 set, the largest proportion of economic benefits is "general", 0.424, followed by "poor", 0.380, indicating that the economic benefits of the project are relatively general at present, The high cost of electricity-hydrogen-storage integrated energy system equipment, resulting in poor profit ratio, so we need to seek technological breakthroughs to reduce system costs and improve economic benefits.

5 Example Analysis

Take the 33-node grid, 6-node hydrogen network as an example, of which the maximum installed capacity of photovoltaic units 9000 KW, the maximum installed capacity of wind turbine 7200 KW, of which the grid 3 nodes are coupled with hydrogen network 2 nodes through P2G device, of which 18 nodes are coupled with hydrogen network 5 nodes through P2G device, wind turbine unit single node grid, photovoltaic unit single node grid as an example (Fig. 1):

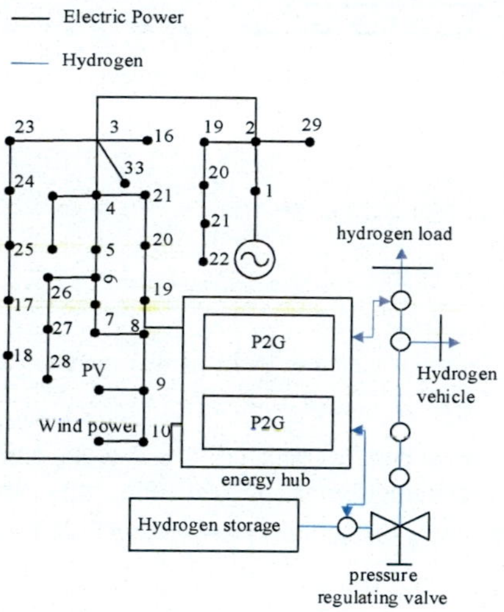

Fig. 1. An integrated energy system architecture diagram

Among them, the output curves of wind power under normal condition and extreme conditions are shown in Fig. 2:

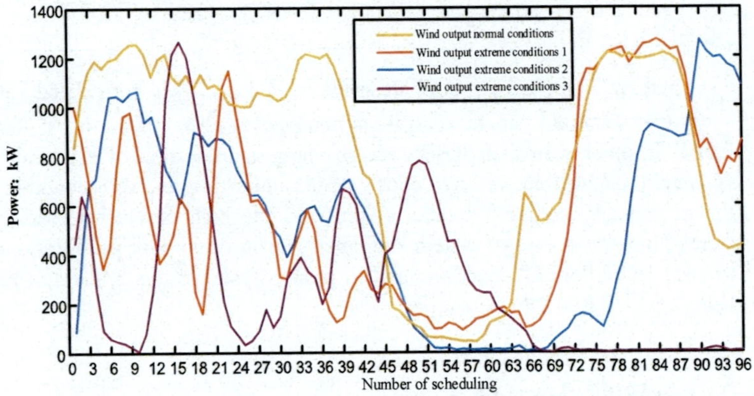

Fig. 2. Wind power under normal condition and extreme conditions

The output curves of photovoltaic power under normal condition and extreme conditions are shown in Fig. 3:

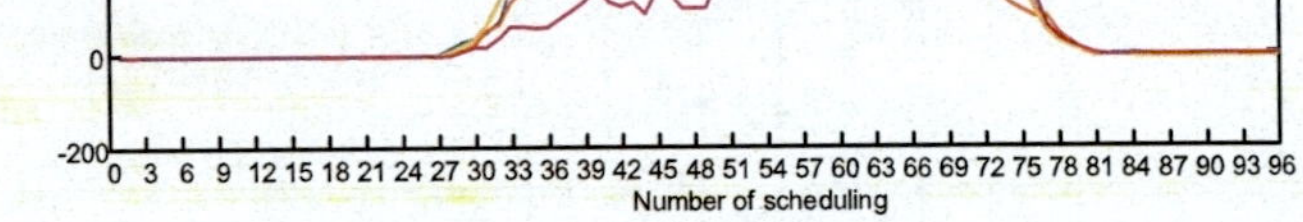

Fig. 3. Photovoltaic power under normal condition and extreme conditions

Comprehensive economic benefits, low-carbon benefits and other six levels of benefit assessment, carbon emission reduction, initial investment in the system and other 20 sub-indicators, the comprehensive evaluation results of the system as shown in Fig. 4:

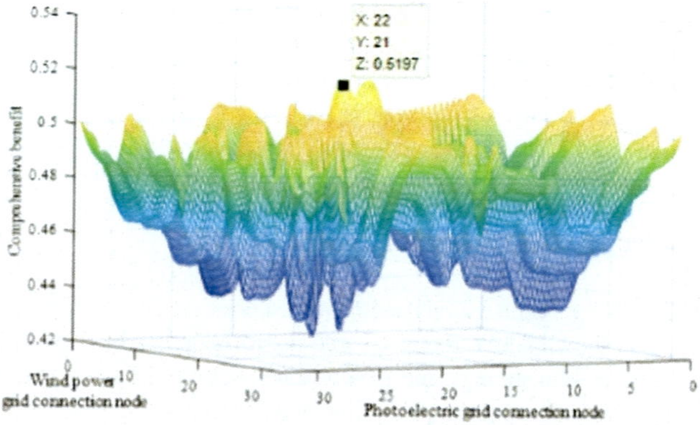

Fig. 4. Comprehensive benefit assessment of IEN

Finally, comprehensive consideration of the various indicators, when the wind turbine at the grid 22 nodes grid, photovoltaic units at the grid 23 nodes grid, in this case, the comprehensive benefits reached the maximum value. In this scenario, the adjacent new energy resources nodes improve the system voltage and reduce the line loss, improve the economic benefit and low-carbon benefit, meanwhile but not because the distance is too close, uncertain output influence each other, resulting in reduced operating benefit.

The comprehensive benefit evaluation of the comprehensive energy system without hydrogen network is carried out. The evaluation process is the same as above. The two results are compared and the comparison of each benefit is obtained in Fig. 5:

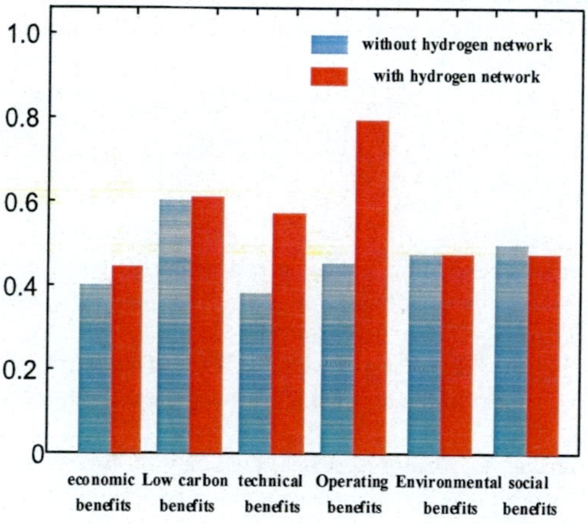

Fig. 5. Comparison of comprehensive benefits of the system with/without hydrogen network

As can be seen from Fig. 5, when the integrated energy system is added to the hydrogen network, it will supplement power generation when the output of new energy is insufficient, and store electric energy when the output of new energy is excessive. Therefore, the stability of the system will be greatly improved, and the operation will be reduced, and the economic benefits will be improved.

6 Summary and Conclusion

Many scholars at home and abroad have explored the comprehensive energy evaluation system, but they mainly focus on economic benefits. In view of the promotion of China's low-carbon strategy and the current situation that a large number of new energy is connected to the power grid, this paper proposes a new comprehensive energy evaluation system considering the uncertain output of new energy:

1) Through analytic hierarchy process, 6 primary indicators and 20 secondary indicators are selected to establish a comprehensive benefit evaluation model.
2) This paper adopts the reverse scenario reduction method and Monte Carlo simulation method to obtain the uncertain scenario of new energy output is obtained.
3) Normalization of related indicators, combined with comprehensive energy assessment model, the final assessment results.
4) The comprehensive benefits of the integrated energy system with hydrogen network and the integrated energy system without hydrogen network are evaluated and the comparative results are obtained.

At present, the treatment of uncertain output mode of new energy needs to be improved, and the carbon emission model needs to be refined. Further research will be conducted in these two aspects in the future.

Acknowledgements. This work is supported by State Grid Jibei Electric Power Company Zhangjiakou Chongli district power supply branch technology project Multi-objective optimal operation for electricity hydrogen storage integrated energy network in Chongli low carbon Winter Olympic Zone (SGJB2JCLF2JS2100095).

References

1. Ma, T., et al.: Energy flow modeling and optimized operation analysis of the microenergy network based on energy hubs. Power Grid Technol. (01), 179–186 (2018). https://doi.org/10.13335/j.1000-3673.pst.2017.0840
2. Su, X.L., Zhao, Z.K., Yang, S., Guo, Y.Q.: Adaptive robust SMC-based AGC auxiliary service control for ESS-integrated PV/wind station. Complexity **2020**(2) (2020). Article ID 8879045, 10 p.
3. Xu, L., Zhang, Y., Zhang, B., Chen, L.: Study on benefit evaluation of integrated energy system based on hybrid multi-attribute group decision method. Ind. Technol. Econ. **33**(03), 52–57 (2014)
4. Jiang, S., Zhang, C., Wu, Y.: Calculation method of natural gas distributed energy station. HVAC **46**(02), 57–60+100 (2016)

5. Dong, F., Zhang, Shang, M.: Multi-Index comprehensive evaluation study of distributed energy system. Chin. Electr. Eng. J. **36**(12), 3214–3223 (2016)
6. Han, Z., Qi, C., Xiang, P., Liu, M., Wang, S.: Benefit analysis and comprehensive evaluation of the distributed energy system. Thermal Power Gener. **47**(02), 31–36 (2018)
7. Chen, H., Huang, H.: Community-level integrated energy system optimization model considering carbon emissions. J. Shanghai Electr. Power Univ. **06**, 613–618 (2020). CNKI: SUN:DYXY.0.2020-06-016
8. Lu, T.: Integrated energy system optimization of carbon emission and renewable energy quota system. Master's thesis, Kunming University of Science and Technology (2021). https://kns.cnki.net/KCMS/detail/detail.aspx?dbname=CMFDTEMP&filename=1021740515.nh
9. Xin, H.: Consider multiple energy complementary clean energy collaborative optimization scheduling and benefit balance research. Doctoral thesis, North China Electric Power University, Beijing (2019). https://kns.cnki.net/KCMS/detail/detail.aspx?dbname=CDFDLAST2020&filename=1019239854.nh
10. Kong Ling State. Research on optimization configuration and coordination and control strategy of scenery hydrogen integrated energy system. Doctoral thesis, North China Electric Power University, Beijing (2017). https://kns.cnki.net/KCMS/detail/detail.aspx?dbname=CDFDLAST2017&filename=1017215717.nh
11. Su, X.L., Chen, L.J., Jun, Y.: On-site engineering test of active support control for the PV station and wind farm in the AC-DC hybrid power grid under extreme fault conditions. Complexity **2021**(1) (2021). Article ID 9636383, 13 p. https://doi.org/10.1155/2021/9636383
12. Li, W.: Study on transmission line path optimization considering the integrated cost. Master's thesis, North China Electric Power University, Beijing (2020). https://kns.cnki.net/KCMS/detail/detail.aspx?dbname=CMFD202101&filename=1021014645.nh
13. Luo, F., Mi, Z., Wang, C., Fang, C., Li, D., Liu, L.: Low-carbon integrated benefit analysis model of grid-connected photovoltaic power generation engineering. Power Syst. Autom. **17**, 163–169 (2014). CNKI:SUN:DLXT.0.2014-17-029
14. Shi, F., Luo, F., Xu, J., Correction: Research on optimal configuration method of multi-region integrated energy distribution system based on planning-operation. Power Syst. Autom. **10**, 125–131 (2020). https://doi.org/10.19635/j.cnki.csu-epsa.000512
15. Cong, H.: Research on the optimal operation of the complementary regional comprehensive energy system considered in the market environment. Doctoral thesis, Shanghai Jiao Tong University (2019). https://kns.cnki.net/KCMS/detail/detail.aspx?dbname=CDFDLAST2020&filename=1020727811.nh
16. Zhang, Y., Li, L., Liu, X.: Comprehensive energy system operating efficiency assessment method of calculating the uncertainty. Renew. Energy **11**, 1671–1678 (2019). https://doi.org/10.13941/j.cnki.21-1469/tk.2019.11.015

Multi-objective Optimal Scheduling for Electricity Hydrogen ESSs Integrated Network in Chongli Winter Olympics Zone

Lize Liu[1], Haijun Liu[2](✉), Shuai Wang[3], Runhao Gao[3], Siwei Liu[3], Yang Gao[4], and Zhengxi Li[5]

[1] Qinghai Key Lab of Efficient Utilization of Clean Energy, Qinghai University, Xining 810016, Qinghai, China

[2] State Key Laboratory of Advanced Power Transmission Technology, Global Energy Interconnection Research Institute Co. Ltd., Beijing, China

[3] Zhangjiakou Chongli District Power Supply Company, State Grid Jibei Electric Power Co., Ltd., Zhangjiakou 075000, China

[4] State Grid Jibei Electric Economic Research Institute, Beijing 100038, China

[5] State Grid Qinghai Electric Power Company Economic and Technological Research Institute, Xining 810016, Qinghai, China

Abstract. With the upcoming 2022 Winter Olympic Games in Beijing. Power system in Chongli has developed significantly because of the increasing share of renewable energy sources (RESs). Despite the RESs, hydrogen stations and energy storage systems (ESSs) are also installed to providing clean energy which forms an integrated network. In order to achieve the goal of low-carbon Olympics and promote RESs consumption, this paper proposed a multi-objective optimal scheduling for electricity hydrogen ESSs integrated network in Chongli Winter Olympics zone. The objective function of the integrated energy network was developed based on operating costs which contains power grid, hydrogen network and ESSs, in addition, carbon emission was taken as the environmental cost. The restrict conditions include hydrogen storage equipment, electric energy storage equipment, RESs, power grid flow, hydrogen network pressure and power flow, and key equipment operation conditions in integrated energy network. The simulation results indicate that the proposed multi-objective optimal scheduling promote RESs consumption in electric-hydrogen-storage integrated energy network. The innovation is to consider the optimization under different hydrogen prices and the optimization under different PV and Wind output, and to consider the influence of the error. The main contribution of this paper is proposing a multi-objective optimal scheduling strategy to adjust the contradiction between supply and demand in electric-hydrogen-storage integrated energy network by arranging various energy systems reasonably under the premise of security constraints. To provide an academic reference to reduce carbon emissions and promote RESs.

Keywords: Integrated energy network · Carbon emissions · Optimal scheduling · Winter olympic games · Hydrogen energy

Z. Hu et al. (Eds.): AIPE 2021, LNDECT 119, pp. 138–149, 2022.
https://doi.org/10.1007/978-3-030-97064-2_14

1 Introduction

Energy crisis is getting worse because of the continuous consumption of fossil energy. In order to adjust the contradictions caused by energy demand and supply, solve problems caused by global environmental deterioration, China has proposed strategies to accelerate its energy transition. For instance, changing the traditional energy dominated structure into an integrated energy network (IEN) mainly depended on renewable energy, or construct multi-energy complementary energy system.

Hydrogen energy plays an indispensable role in the process of energy transformation. (1) Hydrogen energy is the basis of synthetic power fuel and fuel, which has a wide range of applications. For example, hydrogen fuel cell vehicle has excellent environmental protection performance and energy conversion efficiency, which is a significant direction of hydrogen energy utilization in the future. (2) Hydrogen energy can store new energy flexibly according to the supply and demand, and alleviate the problem of balance of electric power and energy caused by the high proportion of new energy connected to the grid. (3) Hydrogen energy is an important element of various energy coupling.

Electric-hydrogen-storage IEN can promote the renewable energy efficient and reduce the irreversible loss of traditional energy as well as complement arity between different energy sources and improve energy efficiency. Therefore, renewable energy and multi-energy IEN are the best answer to the green and long-term stability of human society development.

In Reference [1], the economic optimization scheduling model of the IEN in the park with the total operating cost as the target was constructed, and the scheduling model was processed by using nonlinear problems. Reference [2] has established a unified power flow model of power grid and natural gas network. A unified tidal current model based on Newton descent method is proposed. It is not improved from the nonlinear constraint of natural gas flow and the analysis of heterogeneous energy flow characteristics. Reference [3] has established a risk scheduling model based on line load risk index. A multi-index evaluation method for IEN with integrated transmission and distribution is proposed. A distributed scheduling method for power-gas IEN is proposed, and ADMM iterative method is adopted in algorithm. In reference [4], an improved DC power flow model with graded voltage and reactive power and a warm start iterative solution process are proposed. In reference [5], a hydrogen-natural gas hybrid energy storage system with an intermediate buffer is proposed for power to Gas (P2G) realization system in the electric IEN. A day-ahead optimal scheduling model of energy loss and environmental cost is proposed. In Reference [6], the power flow of transmission network is linearized by an approximate method, a model considering the linear power flow constraint of transmission network is proposed, and an algorithm based on iterative method is established. Reference [7] puts forward the basic architecture and operation strategy of community-level IEN, establishes the objective function of operating cost and environmental cost from two aspects of economy and carbon emission, and uses genetic algorithm to solve. Reference [8] proposed a fixed-point power flow linearization model for three-phase unbalanced network in gas-electricity regional IEN. Under the condition of unified dispatching, a day-ahead optimization multi-time coordinated optimal dispatching model was proposed, giving full play to the dual peak regulating role in gas-electricity regional IEN. Literature [9] proposed a continuous multi-day peak load prediction method based on serial-parallel integrated learning and an energy calculation method suitable for regional IEN containing electricity, heat and gas. Considering two operation modes of grid-connection and isolated island, the Jacobian matrix reflecting energy coupling relationship was obtained. In reference [10], the dynamic model of solar PV hydrogen IEN was established, and the optimal configuration of solar PV hydrogen IEN with Wind and PV abandonment was considered, and an improved chemical reaction optimization algorithm was proposed to maximize profits. Reference [11, 12] uses electric vehicles to participate in high permeability photoelectric and Wind power consumption to reduce microgrid cost, Reference [13, 14] introduces joint industries such as photothermal power station and hot and cold power supply to guide user electricity behavior with retail electricity price and retail heat price, Reference [15] classification models the user load of micro-grid community and guide residential electricity consumption with time-sharing electricity price.

The rest of this paper is organized as follows: Sect. 2, introduce the architecture of IEN and its operating cost and environmental cost. Section 3, the objective function is constructed with economic cost and environmental cost as the target. Considering the grid power flow constraints, hydrogen network power flow constraints and energy storage constraints, a multi-objective optimal power flow algorithm with the minimum economic cost and environmental cost is presented in Sect. 3, Sect. 4, taking Chongli as an example, the power flow algorithm is applied to the day-ahead optimization dispatching in Chongli area to ensure the economic and environmental protection operation of the electric-hydrogen-storage IEN.

2 IEN Architecture

Wind farms and PV stations are connected to hydrogen energy storage devices, and provide high quality power to grid through the energy management system, which are used for power supply and hydrogen production. When the power generation in IEN is greater than load demand, the surplus power is converted into hydrogen energy and stored by hydrogen energy storage equipment. Otherwise, the energy management system starts the fuel cell to supplement the power shortage.

3 Objective Function

The overall operating cost of electric-hydrogen-storage IEN is taken as the lowest target, which consist operating cost and environmental cost, the objective function is as follows

$$\min F = \alpha F_1 + \beta F_2 \tag{1}$$

Where F is the comprehensive cost of IEN, F_1 is the operating costs of IEN, F_2 is the environmental costs of IEN, a is the operating cost coefficient, β is the environmental cost coefficient of.

3.1 IEN Operating Costs

The electric-hydrogen-storage IEN operating cost is

$$F_1 = F_{WT} + F_{PV} + F_W + F_G + F_H + F_{SH} \tag{2}$$

Where F_{WT} and F_{PV} are the conventional operating costs of Wind power and PV units respectively. F_W and F_G are wind and PV abandoning penalty costs. F_H is the hydrogen production energy storage unit operating cost, F_{SH} represent energy loss in IEN.

3.1.1 Operation Cost of Wind and PV Unit

Equation (3) gives the operation cost of wind and PV unit

$$\begin{cases} F_{WT} = \sum_{t=1}^{T}\sum_{i=1}^{N_{WT}} u_{i,t}(a_i P_{i,t} + b_i P_{i,t} + c_i) \\ F_{PV} = \sum_{t=1}^{T}\sum_{i=1}^{N_{PV}} u_{i,t}(a_i P_{i,t} + b_i P_{i,t} + c_i) \end{cases} \tag{3}$$

Where T is the total amount of time periods in the scheduling period. N_{WT} is the number of Wind turbines, N_{PV} is the number of PV units, $u_{i,t}$ is a variable of 0–1, indicating the start-stop state of unit i at t, a_i, b_i and c_i are cost coefficients of unit i, $P_{i,t}$ is the output power of unit i at t.

3.1.2 Punishment Cost of Abandoning Wind and PV

$$\begin{cases} F_W = \sum_{t=1}^{T} C_W(P'_{Wt} - P_{Wt}) \\ F_G = \sum_{t=1}^{T} C_G(P'_{Gt} - P_{Gt}) \end{cases} \tag{4}$$

Where C_W and C_G are wind abandoning and PV abandoning penalty cost coefficients respectively, P'_{Wt} and P'_{Gt} are respectively the predicted output values of wind farm and PV field in time period t. P_{Wt} and P_{Gt} are the actual output values of Wind farm and PV field in time period t respectively.

3.1.3 Operating Cost of Hydrogen Energy Storage

$$F_H = \sum_{t=1}^{T} \Delta t \delta_t \left\{ (1 - \partial_{E2H}) \sum_{i=1}^{N_{EL}} P_{i,t}^{EL} + (\frac{1 - \beta_{H2E}}{\beta_{H2E}}) \sum_{i=1}^{N_{FC}} P_{i,t}^{FC} \right\} \tag{5}$$

Where Δt is the Time step, δ_t is the price of power grid at the first moment, α_{E2H} is the efficiency of electrolytic hydrogen production, N_{EL} is the total number of system electrolytic cells, $P_{i,t}^{EL}$ is the active power consumed by the electrolytic cell at any moment, β_{E2H} is the Fuel cell power generation efficiency, N_{FC} is the Total number of fuel cells, $P_{i,t}^{FC}$ is the Contributes to the active power of the instantaneous fuel cell.

3.1.4 Line Loss Cost

$$\begin{cases} F_{SH} = \Delta P_1 + \Delta P_2 \\ \Delta P_1 = \dfrac{(P_L - P_{DG})^2 + (Q_L - Q_{DG})^2}{U}(R + jX)L \\ \Delta P_2 = \dfrac{(P_L + Q_L)^2}{U}(R + jX)(M - L) \end{cases} \quad (6)$$

Where ΔP_1 are ΔP_2 respectively the line loss between the end bus to the distributed power supply and the distributed power supply to the load, P_L and Q_L are the active and reactive power of the load respectively. P_{DG} and Q_{DG} are respectively the active and reactive power of the power supply. M is the distance from power supply to load, L is the distance from power supply to distributed power supply.

3.2 Environmental Cost of IEN

In the electric-hydrogen-storage IEN, carbon emissions are taken as the environmental cost, and the environmental cost is expressed as follows:

$$\begin{cases} F_2 = E_D \eta_t N - E_{MG} \eta_T Z \\ E_{MG} = E_D + E_N \end{cases} \quad (7)$$

Where E_D is the CO_2 emission during system power generation, η_t is the proportion coefficient of carbon tax, N is the tax payable per unit of carbon emission, E_{MG} is the sum of CO_2 emissions from power generation of the electric-hydrogen-storage IEN and CO_2 emissions from power purchase from the grid, η_T is the proportion of CO_2 emission reduction in total emissions, Z is the transaction price, E_N is the CO_2 emissions from purchasing electricity from the grid.

4 The Constraint Conditions

4.1 Constraints on Power Grid

4.1.1 New Energy Primary Resource Constraint

$$\begin{cases} 0 \le P_{WP}(t) \le P_{\max}^{WP} \\ 0 \le P_{PV}(t) \le P_{\max}^{PV} \end{cases} \quad (8)$$

Where $P_{WP}^{\max}, P_{PV}^{\max}$ is the are the maximum Wind power output and PV output in Chongli region respectively.

4.1.2 Power Flow Constraint of Power Grid

(1) Power balance

$$\begin{cases} P(i,t) = \sum_{j=1}^{nbus} (G_{ij}U(i,t)U(j,t)\cos(\theta_{ij}(t)) + B_{ij}U(i,t)U(j,t)\sin(\theta_{ij}(t))) \\ Q(i,t) = -\sum_{j=1}^{nbus} (B_{ij}U(i,t)U(j,t)\cos(\theta_{ij}(t)) - G_{ij}U(i,t)U(j,t)\sin(\theta_{ij}(t))) \end{cases} \quad (9)$$

Where $P(i, t)$, $Q(i, t)$ is the represents the active power and reactive power of the moment node respectively, G_{ij}, B_{ij} are the real and imaginary parts of the row and column in the node admittance matrix respectively, $U(i, t)$ is the Voltage amplitude of the moment node, $\theta_{ij}(i)$ is the phase Angle difference at both ends of the moment branch, *nbus* is the number of system nodes.

(2) Node voltage amplitude constraint

$$U_{\min}(i) \leq U(i,t) \leq U_{\max}(i) \quad (10)$$

Where $U_{\max}(i)$, $U_{\min}(i)$ are the upper limit and lower limit of voltage amplitude allowed at each node.

4.2 Hydrogen Network Constraints

4.2.1 Hydrogen Storage Constraints of Hydrogen Energy Storage Equipment

For hydrogen energy storage device, the constraint conditions of hydrogen energy storage must be

$$S_{OC,\min} \leq S_{OC} \leq S_{OC,\max} \quad (11)$$

Where $S_{OC,\min}$ is the minimum load state of hydrogen energy storage equipment, $S_{OC,\max}$ is the maximum load state of hydrogen energy storage equipment, S_{OC} is the charged state of the energy storage device at a certain moment.

4.2.2 Constraints of Hydrogen Network Air Pressure and Power Flow

(1) The pressure constraint of hydrogen net is

$$H_i^{\min} \leq H_{i,t} \leq H_i^{\max} \forall i \in N_g, \forall t \in T \quad (12)$$

Where $H_{i,t}$ is the air pressure of node i at time t, $H_i^{\min}$ and $H_i^{\max}$ are the upper and lower pressure of the node respectively.

(2) The power flow constraint of hydrogen network is:

$$Q_{ij,t}^{gas} = K_{ij}S_{ij,t}\sqrt{S_{ij,t}\left(\left(p_{i,t}^{gas}\right)^2 - \left(p_{j,t}^{gas}\right)^2\right)}, \forall t, \forall ij \in E^{gas} \tag{13}$$

Where $Q_{ij,t}^{gas}$ is the represents the flow rate of hydrogen pipeline, $p_{i,t}^{gas}$, $p_{j,t}^{gas}$ are pressures upstream and downstream of the pipeline respectively, K_{ij} is the comprehensive coefficient of pipe and hydrogen fluid, E_{gas} is the collection of gas network branches, $G_{ij,t}$ is the gas flow direction of pipeline at moment.

5 Analysis of Examples

The IEN shown in Fig. 1 is taken as an example to verify the proposed strategy with 24 h scheduling cycle and 15 min scheduling time. The grid node 3 connect to hydrogen network node 2, grid node 18 connect to and hydrogen network 5 node for the model, select 25 Wind node, selection of 27 PV node for simulation analysis.

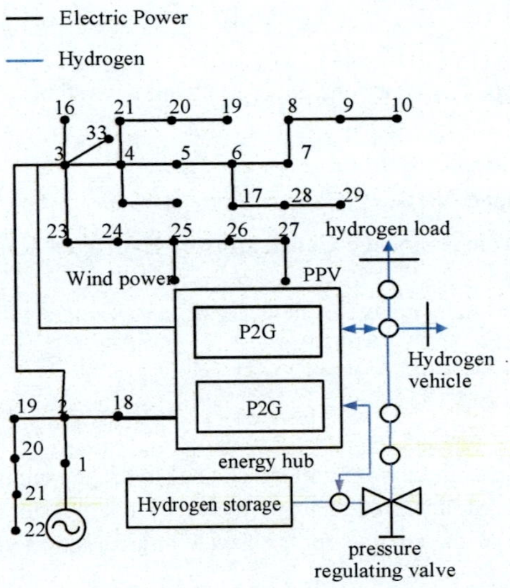

Fig. 1. Electricity hydrogen ESSs integrated network

5.1 Basic Data

Taking the IEN of electricity, hydrogen and storage in Chongli Low-carbon Olympic Zone as an example, the dispatching period is one day (24 h), and the unit dispatching time is 15 min. In this paper, four different wind power outputs are used for optimization scheduling example analysis, and wind power outputs are shown in Figs. 2 and 3. Table 1 lists the power purchase and sale prices of the system in different periods.

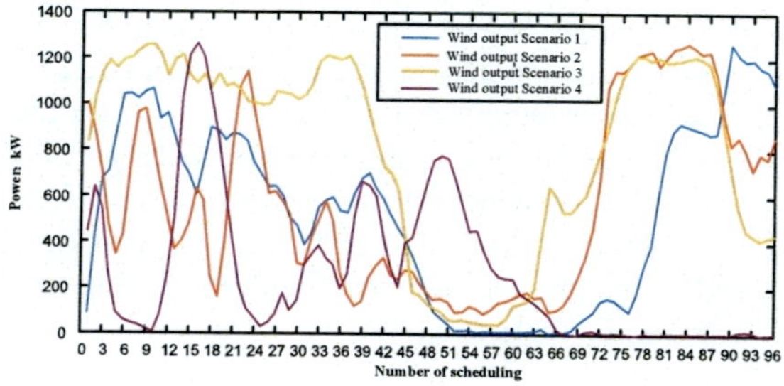

Fig. 2. Wind output curves

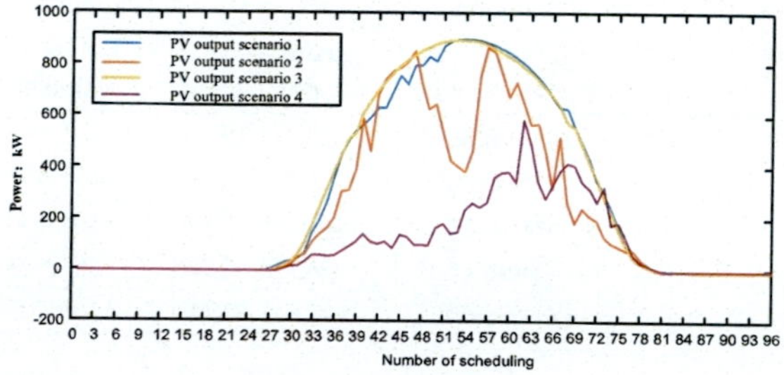

Fig. 3. PV output curves

Table 1. Power purchase and sale prices in different periods of time

Way to trade	The trough time	The ordinary time	The peak time
Purchase prices	0.45	0.65	0.85
Sale prices	0.40	0.55	0.75

5.2 Single Objective Optimization

5.2.1 Comparison of Operating Costs and Environmental Costs Under Different hydrogen Prices

The operation cost of this paper directly considers the sum of electricity purchase cost and hydrogen purchase cost. By comparing the price of electricity and hydrogen in each period, the lowest price is selected to achieve the minimum operation cost. In this paper, the load of each period is subtracted from the electricity generated by solar output. This paper takes hydrogen price 0.6048 (yuan/kW.h) as the standard value, and forms an obvious contrast between environmental cost and operating cost by comparing different hydrogen prices. Fan output scenario 1 and PV output scenario 1 are selected as the total output of the operating cost and environmental cost of different hydrogen prices of the electric-hydrogen-storage IEN, and the operating cost and environmental cost of other different outputs can refer to this result.

The carbon emission is taken as the environmental cost, and the carbon emission of Wind power generation and PV power generation is 0 in this paper. Since the energy storage device is hydrogen energy storage, that is, the carbon emission is 0, so the CO_2 generated by power grid purchase is taken as the environmental cost. The tax required for CO_2 per unit carbon emission is 50 yuan/t, and the tax ratio coefficient is 100%. The CO_2 emission caused by one degree of electricity is 0.96 kg.

In this paper, the economic and environmental costs before and after optimization are selected in the scenic-output scenario. Other situations can refer to this result. The economic and environmental costs under different hydrogen prices in scenic-output scenario are shown in Table 2.

Table 2. Optimization comparison of economic cost and environmental cost under different hydrogen prices

Hydrogen prices (yuan/kW.h)	Economic cost/yuan		Environmental cost/yuan	
	Before optimization	After optimization	Before optimization	After optimization
0.45	48096.6625	23397.4891	3465.4031	0
0.5048	48096.6625	35337.4479	3465.4031	969.6709
0.6048	48096.6625	40536.8899	3465.4031	969.6709
0.7048	48096.6625	44314.4749	3465.4031	2215.0931
0.85	48096.6625	25955.7571	3465.4031	2215.0931

5.2.2 Optimization Comparison of Operating Cost and Environmental Cost in Different PV and Wind Output Scenarios

This paper takes hydrogen price 0.6048 (yuan/kW.h) as the standard value to compare the operating cost and environmental cost in different PV and Wind output scenarios. In other cases, the operating cost and environmental cost of different scenic-output scenarios with different hydrogen prices can refer to this result. Table 3 shows the comparison of operating cost and environmental cost before and after optimization with hydrogen price 0.6048 (yuan/kW.h) of different PV and Wind output.

Table 3. Optimization comparison of operating cost and environmental cost under different PV and Wind outputs

Output in different PV and wind	Economic cost/yuan		Environmental cost/yuan	
	Before optimization	After optimization	Before optimization	After optimization
PV and wind output one	48096.6625	40536.8899	3465.4031	969.6709
PV and wind output two	47930.3168	40569.5025	3541.5344	1257.0019
PV and wind output three	31009.2247	25676.6208	2160.1758	476.7935
PV and wind output four	77079.5036	64108.3650	5505.8674	1632.7331

5.3 Multi-objective Optimization

In this paper, the stable output of PV and Wind is selected to compare the operating cost and environmental cost of different output under different hydrogen prices, because the stable output can better carry out multi-objective optimization, and other output can refer to the stable output. Choose PV and Wind output three as a stable output. The selection range of hydrogen price is 0.45–0.85. In order to observe the result after optimization more intuitively, compared the reduced operating cost and environmental cost after optimization to make a more intuitive description. The reduced operating cost is shown in Fig. 4, and the reduced environmental cost is shown in Fig. 5.

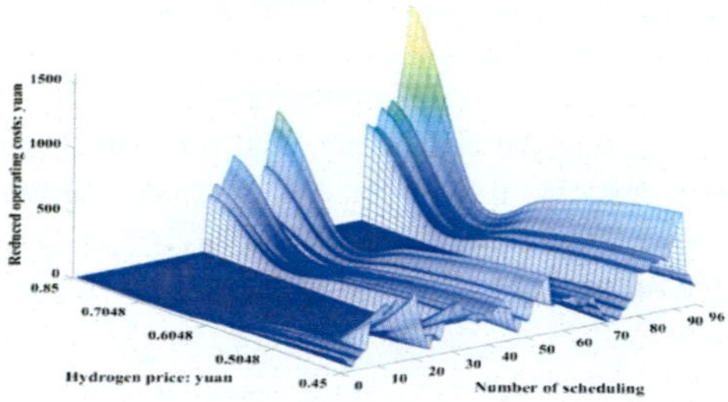

Fig. 4. Optimization model of operation cost reduction

According to the Fig. 4, the reduced operating cost of the model is 20–23 points at 10 and 12–14 points, because the electricity price reaches the peak price in the region at 10–14 points, 20 and 20–23 points, while the hydrogen price ranges from 0.45–0.85, at 10–14 points, 20–23 points, so the electricity-hydrogen-storage system will shop hydrogen to hydrogen. Moreover, the Fig. knows that the lower the hydrogen price, the higher the reduced operating cost.

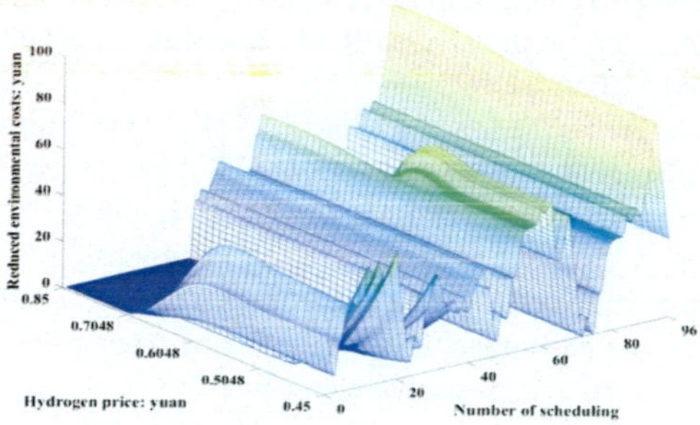

Fig. 5. Optimization model of environmental cost reduction

According to the Fig. 5, at 10 points–12 points, 12 points–14 points, 20 points–23 points the reduction of operating cost is higher, which is because the hydrogen price is lower than the electricity price, so hydrogen shopping to hydrogen, will not produce CO_2.

5.4 Example of the Analysis Results

With the above examples, we derive the following results:

1) By comparing the output of different hydrogen prices in the same PV and Wind, it is concluded that the lower the hydrogen price, the lower the operating cost, and the environmental cost will also decrease.
2) By comparing the situation of different output at the same hydrogen price, the optimized operation cost and environment cost of the electric-hydrogen-storage IEN are stable, and the optimized operation cost and environment cost in the electric-hydrogen-storage IEN are the maximum.
3) Through comparing the error analysis of different output, the operation cost and environment cost optimized in the electric-hydrogen-storage IEN is stable, and the operation cost and environment cost optimized in the electric-hydrogen-storage IEN are the maximum.

6 Summary and Prospect

In order to achieve the goal of low-carbon Olympics and promote RESs consumption, this paper proposed a multi-objective optimal scheduling for electricity hydrogen ESSs integrated network in Chongli Winter Olympics zone:

1) This paper introduces the characteristics of fossil energy and new energy and the role of hydrogen energy and electric hydrogen storage IEN in human life, and the IEN structure of electric hydrogen storage is established.
2) In the case of electric hydrogen storage IEN architecture, an objective function based on operating cost and environmental cost is established. The current trend nonlinear constraints in the power grid and the hydrogen network are converted into linear constraints through the section selection strategy and the segmented linearization of the iterative method. A scheduling scheme based on the minimal objective function is obtained.
3) The results are verified by the 33 node of electric hydrogen storage IEN in Chongli area of Zhangjiakou. The proposed multi-objective optimal scheduling promotes RESs consumption in electric-hydrogen-storage IEN.
4) The main contribution of this paper is proposing a multi-objective optimal scheduling strategy to adjust the contradiction between supply and demand in electric-hydrogen-storage IEN by arranging various energy systems reasonably under the premise of security constraints.

At present, in the process of converting the hydrogen gas into the electric energy, the efficiency of the fuel cell is too low, so that the hydrogen gas in the process of converting the electric energy, the energy is lost in the form of the thermal energy. How to improve the performance of the fuel cells is still the focus of the future research.

Acknowledgments. This work is supported by State Grid Jibei Electric Power Company Zhangjiakou Chongli district power supply branch technology paper Multi-objective optimal operation for electricity hydrogen storage IEN in Chongli low carbon Winter Olympic Zone (SGJB2JCLF2JS2100095).

References

1. Sun, Q., Xie, D., Nie, Q., Zhang, L., Chen, Q., Chen, J.: Research on economic optimization scheduling of IEN with electric-heat-cold-gas load in campus. China Electr. Power **53**(04), 79–88 (2020)
2. Su, X.L., Zhao, Z.K., Yang, S., Guo, Y.Q.: Adaptive robust SMC-based AGC auxiliary service control for ESS-integrated PV/wind station. Complexity **2020**(2) (2020). 10 p. Article ID 8879045
3. He, Y.: Research on stochastic optimization and distributed scheduling of power-gas IEN. Zhejiang University (2019)
4. Fan, Z., Zhu, J., Yuan, Y., Wu, H.: Distributed power planning model for active distribution network based on improved DC power flow algorithm and its linearization method. Power Grid Technol. **43**(02), 504–513 (2019)
5. Liu, J., Zhou, C., Gao, H., Guo, Y., Zhu, Y.: Optimization of day-ahead economic dispatching of electric-gas integrated energy microgrid considering hydrogen and natural gas hybrid energy storage. Power Grid Technol. **42**(01), 170–179 (2018)
6. Zhu, J., Shi, K., Li, Q.: Timing production simulation and new energy consumption capacity evaluation considering power flow constraint of transmission network. Grid Technol. 1–9 (2021). https://doi.org/10.13335/j.1000-3673.pst.2021.0757
7. Hua, C., Haiye, H.: Optimization model of community-level IEN considering carbon emissions. J. Shanghai Dianji Univ. **36**(06), 613–618 (2020)
8. Jiang, Y.: Research on Coordinated and Optimized Operation Strategy of Gas-Electricity Regional IEN, Wuhan University, p. 124 (2019)
9. Shi, J.: Research on Supply and demand prediction and optimal operation technology of regional IEN, North China Electric Power University, Beijing, p. 136 (2019)
10. Lingguo, K.: Optimal configuration and coordinated control strategy of solar hydrogen IEN, North China Electric Power University, Beijing, p. 156 (2017)
11. Wang, J., Shi, J., Wen, T., et al.: Optimization operation of cogeneration micro-network for demand response. Power Syst. Autom. **43**(1), 176–185 (2019). https://doi.org/10.7500/AEPS20180216001
12. Cheng, S., Huang, T., Wei, R.: Multi-time scale optimization scheduling of hot-cold and electricity supply micro network with ice storage air conditioning. Autom. Power Syst. **43**(5), 30–38 (2019)
13. Su, X.L., Chen, L.J., Jun, Y.: On-site engineering test of active support control for the PV station and wind farm in the AC-DC hybrid power grid under extreme fault conditions. Complexity **2021**(1) (2021), 13 p. Article ID 9636383
14. Nan, S., Zhou, M., Li, G.: Optimal residential community demand response scheduling in smart grid. Appl. Energy **210**, 1280–1289 (2018)
15. Cheng, X., Li, J., Cao, L., Liu, X.: Distributed parallelization algorithm for the optimal tidal flow of a power system. Power Syst. Autom. **24**, 23–27 (2003)

The Evaluation Index System of Electric Power Industry Supply Chain Management Based on FAHP

Xia Zhenlai[1], Zou Ruyi[1], Zhang Guanxiang[2](✉), and Zhong Huiling[2]

[1] Energy Development Research Institute, China Southern Power Grid, Guangzhou, China
[2] Department of Electronic Commerce, South China University of Technology, Guangzhou 510006, China
gxzhang@scut.edu.cn

Abstract. The development of modern supply chain has continuously put forward requirements for the reform of electric power industry supply chain management. Based on modern supply chain development requirements and electric power industry supply chain demand characteristics, this paper constructs the electric power industry supply chain management QBENS evaluation index system from five dimensions: quality, benefit, efficiency, norm, and service, and uses the fuzzy analytic hierarchy process (FAHP) to determine the evaluation index weight. Different from other research literature, this index system can comprehensively reflect the performance level of electric power industry supply chain management, and include the characteristics of digital supply chain and the development concept of green supply chain. Finally, the effectiveness and feasibility of the evaluation index system are verified through the case study of a large power grid company in China.

Keywords: Electric power industry · Modern supply chain · Fuzzy analytic hierarchy process · Evaluation index system

1 Introduction

With the continuous development and upgrade of the modern supply chain, electric power companies increasingly require the supply chain management system to be scientific, systematic and coordinated [1]. The supply chain management evaluation index system is a key means to evaluate and analyze the supply chain management and development. The evaluation content of the electric power industry supply chain management evaluation index system should meet the demand characteristics of the electric power industry supply chain, be scientific and effective, and be consistent with the development requirements of the modern supply chain, that is, highlighting the characteristics of planning, transparency, real-time, collaboration, and intelligence of the digital supply chain [2], including advanced environmental protection concepts

Z. Hu et al. (Eds.): AIPE 2021, LNDECT 119, pp. 150–159, 2022.
https://doi.org/10.1007/978-3-030-97064-2_15

such as green supply chain for resource conservation and environmental protection [3], in order to play the guiding role of supply chain management evaluation. Therefore, it is essential to establish an evaluation index system suitable for modern supply chain management in the electric power industry.

At present, the research on the evaluation of supply chain management in the electric power industry is mainly from a partial perspective, such as supplier performance evaluation [4], supply chain risk evaluation [5], supply chain benefit evaluation [6], power equipment full life cycle assessment [7], supply chain green evaluation [8], etc., which do not cover all business activities in the electric power industry supply chain. The research literature on the comprehensive evaluation of the entire supply chain business management of the electric power industry is still relatively limited. That is, reference [9] started from the material management evaluation index and the comprehensive enterprise management evaluation index, and established a power grid enterprise material management level evaluation index system based on combined weights, but it did not pay enough attention to the concept of digitalization and green supply chain. Reference [10] started from the connotation and characteristics of the power grid material supply chain, and designed an evaluation index system with strategic management, service quality, operational efficiency, operational norms, operational efficiency, and innovative development as the first-level indicators. But it lacks case analysis and indicators have not yet been quantified. There are also many systematic evaluation studies on the entire supply chain business management in other industries, but the applicability to the electric power industry is not high.

There are many mature research theoretical foundations for the construction of supply chain management evaluation index system, such as the Supply Chain Balanced Scorecard Model (BSC) proposed by Kaplan and Norton in 1992 is guided by strategic management and is an evaluation model that takes into account the performance of supply chain business processes and the development of corporate strategy [11]. The four aspects of the internal operation of the supply chain, customers, finance, and future innovation and growth have systematically evaluated the business development of the company, and proved in practice that it can help the company to better conduct overall management. The Supply Chain Operation Reference Model (SCOR) proposed by the Supply Chain Management Association in 1996 is based on process management, constructing a multi-functional supply chain performance evaluation system, and its evaluation indicators are divided into five dimensions: reliability, responsiveness, agility, cost and asset management. It has strong operability [12]. The Supply Chain Maturity Model (DDVN) was proposed by Gartner in 2003. It is demand-driven and focuses on the qualitative evaluation of supply chain maturity, mainly from strategy and organization, product life cycle, and supply network design. Design evaluation index system on 7 basic dimensions including demand management, supply operation, customer satisfaction and supply chain management, analyze the management performance of the supply chain and the efficiency, benefit and effectiveness brought by management investment, and position the enterprise development stage [13].

The QCDS evaluation model was originally proposed by Toyota as one of the indicators to measure the supply level of its suppliers, that is, to evaluate the performance of manufacturing companies' parts suppliers from the four aspects of quality, cost, delivery, and service, in order to improve the quality management of enterprises [14]. The above models all mainly focus on manufacturing enterprises. Based on the SCOR model, the Supply Chain Committee of China Electronic Commerce Association proposed a reference model for supply chain management performance evaluation (SCPR) in 2003, which focuses on the field of commercial circulation, focusing on the five aspects of internal supply chain order response ability, customer satisfaction, business standard coordination, node network effect and system adaptability, but it lacks the evaluation of procurement and coordination of procurement and sales [15].

Based on the existing research, this paper proposes a QBENS evaluation index system for supply chain management in the electric power industry. In a different way to the foregoing literature, this evaluation index system includes advanced concepts of modern supply chain such as digital supply chain and green supply chain, and comprehensively describes the overall operation and development level of the electric power industry supply chain through the evaluation of the five dimensions of quality, efficiency, benefit, norm and service. The index weight is determined by FAHP, which ensures the scientific and rationality of the index weight, and a variety of scoring methods are used to score the indicators. Through case study, it can be seen that the evaluation index system can help electric power industry managers find problems in the operation of the supply chain, improve them in a timely manner, and guide the supply chain to develop in a better direction.

2 QBENS-Based Electric Power Industry Supply Chain Management Evaluation Index System

Before constructing the electric power industry supply chain management evaluation index system, firstly based on the supply chain operation reference model (SCOR) to sort out the business process of the power industry supply chain. It can be seen that the electric power industry supply chain covers demand planning, procurement management, supplier management, quality control management, warehousing and distribution, reverse logistics, supply chain supervision and other business links. And based on supply chain evaluation models such as the supply chain balanced scorecard model (BSC), supply chain maturity model (DDVN), QCDS evaluation model, supply chain management performance evaluation reference model (SCPR), it is preliminarily determined that the construction dimensions of the supply chain management evaluation index system need to include quality management, financial performance, customer service, business management, development potential and so on.

In addition, the electric power industry supply chain has unique demand characteristics compared with other industry supply chains. The electric power industry supply chain management focuses on excellent material quality, effective financial resources, efficient operation and maintenance and emergency protection. In the actual operation, it pays attention to the supporting role of the supply chain in resource guarantee, value creation, digital operation and risk control. Based on the above, the QBENS evaluation index system is proposed to evaluate the electric power industry supply chain management from the five dimensions of quality, benefit, efficiency, norm, and service. In each dimension, it should select key typical and highly representative indicators to comprehensively reflect the key business operation level of electric power industry supply chain, highlight the inheritance and development of the existing core indicators of electric power industry supply chain management, and ensure the continuity of management to a certain extent. In addition, it must follow the indicator construction principles of science, objectivity, rationality, foresight, advancement, and maneuverability to ensure the logical unity and normal implementation of indicators. Among them:

The quality dimension (Q) mainly evaluates the level of material and quality management in the electric power industry, reflects the supporting role of resource guarantee at the strategic level, and reflects the service guarantee capability of the entire supply chain.

The benefit dimension (B) mainly examines the cost of the supply chain and financial-related process links of the electric power industry, reflects the level of value creation at the strategic level, and reflects the full-cycle management capability of the supply chain.

The efficiency dimension (E) mainly measures whether the operation of each link in the electric power industry supply chain is efficient, reflects the effectiveness of digital operations at the strategic level, and reflects the ability to control the efficiency of all elements of the supply chain.

The normative dimension (N) mainly evaluates the compliance of the various processes of the electric power industry supply chain and the social responsibilities assumed, reflects the status of supply chain risk control, and reflects the ability of the supply chain to prevent and control all-round risks.

The service dimension (S) mainly examines the satisfaction of internal customers in the supply chain of the electric power industry with the supply of materials, as well as the satisfaction of cooperation with external partners, reflecting the comprehensive service level of the supply chain.

Starting from five dimensions, construct the electric power industry supply chain evaluation index system as shown in the Fig. 1, which contains 32 indicators, all certified by experts to ensure their applicability.

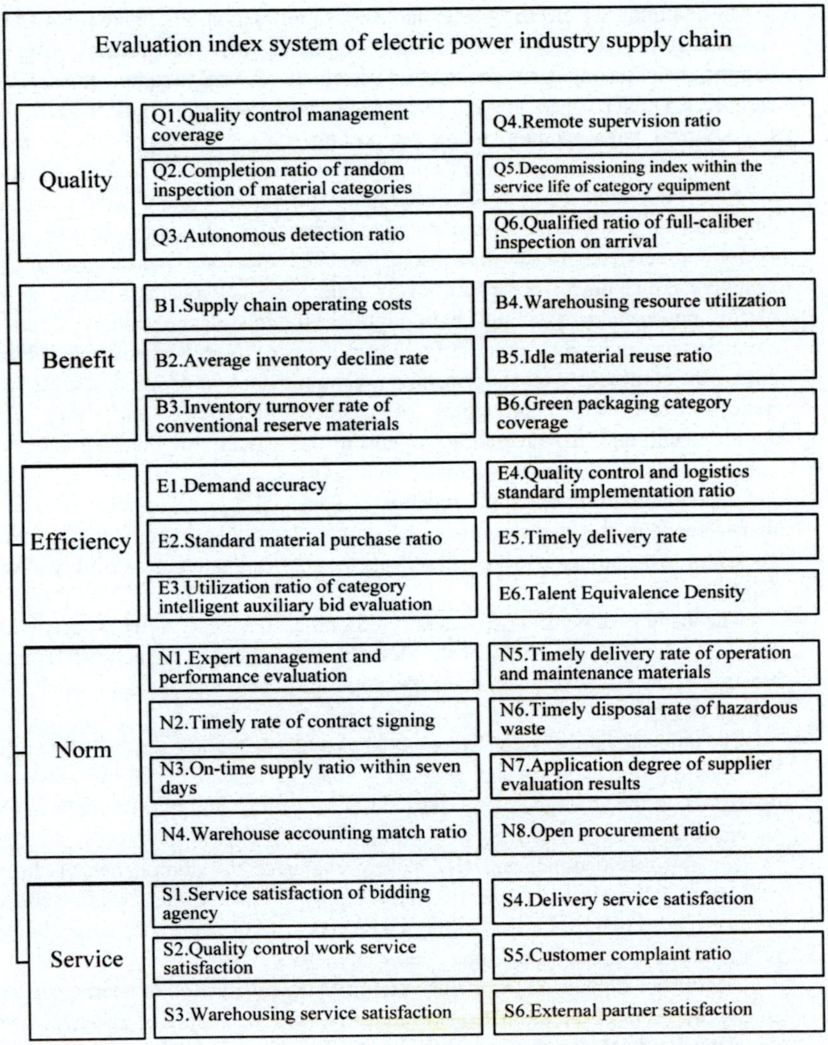

Fig. 1. QBENS evaluation index system for electric power industry supply chain management

The evaluation index system completely covers the main activities of supply chain management in the electric power industry, comprehensively reflects the overall operation level of the supply chain, includes the concept of green supply chain (e.g. green packaging category coverage), considers the planning, transparency, real-time, coordination, wisdom and other characteristics of digital supply chain (e.g. remote supervision ratio), and gives full play to the guiding role of supply chain management evaluation.

3 Evaluation Method Based on FAHP

This paper uses fuzzy analytic hierarchy process (FAHP) to determine the index weight. Analytic hierarchy process is a decision-making method that decomposes the elements related to the evaluation object into objectives, criteria and schemes, and carries out qualitative and quantitative analysis on this basis. In the analytic hierarchy process, the complex index system is decomposed into classes, the hierarchical structure model is established, and the importance comparison between indexes is used to determine the weight of indexes. FAHP uses fuzzy consistent judgment matrix to compare the importance of each index, and improves the consistency test and weight calculation of analytic hierarchy process judgment matrix [16]. FAHP can solve the problems of uncertainty and mental decision-making problems. This approach is a type of decision-making method that can determine the weight of a data group [17].

The fuzzy analytic hierarchy process determines the index weight according to the following steps.

3.1 Construction of the Fuzzy Consensus Decision Matrix

The fuzzy consensus judgment matrix R is an $n \times n$ matrix. It is obtained by expert scoring and other methods and obtained through the opinions of statistical experts.

$$R = \begin{bmatrix} r_{11} & r_{12} & \cdots & r_{1n} \\ r_{21} & r_{22} & \cdots & r_{2n} \\ \vdots & \vdots & \vdots & \vdots \\ r_{n1} & r_{n2} & \cdots & r_{nn} \end{bmatrix} \quad (1)$$

Its elements r_{ij} have the following properties:

$$\begin{cases} r_{ii} = 0.5 \\ r_{ij} = 1 - r_{ji} \\ r_{ij} = r_{ik} - r_{jk} + 0.5^{,(i,j,k=1,2,\ldots,n)} \\ 0 \leq r_{ij} \leq 1 \end{cases} \quad (2)$$

In the formula, r_{ij} represents the relationship between the *i-th* index and the *j-th* index. A value of 0.5 means that the two indexes are equally important. The larger the value, the more important the index *i*.

3.2 Calculation of the Weight of Each Secondary Index

The weight of each index of the fuzzy analytic hierarchy process is obtained by matrix calculation, so as to determine the relationship between the hierarchical indexes and calculate the higher-level indexes. The calculation method of the secondary index weight w_i is:

$$w_i = \frac{\beta^{\frac{1}{n}\sum_{j=1}^{n} r_{ij}}}{\sum_{k=1}^{n} \beta^{\frac{1}{n}\sum_{j=1}^{n} r_{kj}}}, i = 1, 2, \ldots, n \quad (3)$$

Where n is the number of index. β is the parameter that determines the weight resolution. This paper takes the resolution parameter $\beta = e^3$. According to the above calculation method, the weight of the whole index system can be obtained.

3.3 Determination of Comprehensive Evaluation Score

Due to the different development scale, goals and characteristics of different electric power enterprises in different regions, the indicators cannot be compared horizontally. According to the principle of combining horizontal and vertical comparisons, this paper uses different scoring methods for different secondary indicators, such as interval distribution method, index compliance method, increase and decrease comparison method [18]. In addition, the index system uses a combination of quantitative and qualitative indicators to improve fairness, rationality and objectivity of the result [19]. The scores of the secondary indicators are independent of each other.

The dimension score b_d synthesis is carried out by multiplying and adding operator M (•, ⊕) through secondary indicators of each dimension.

$$b_d = \sum_{x=1}^{n} w_i r_{dx} \quad (4)$$

In the formula, d is one of the five dimensions (QBENS), and r_{dx} represents the score of the *x-th* secondary index under the dimension *d*.

4 Case Study

4.1 Case Selection

This paper selected two branches of a large power grid enterprise in regions A and B as the analysis objects, collected the data in the first quarter of 2021, and used the QBENS evaluation index system and the above evaluation methods to evaluate the operation of the power grid supply chain in the two regions. Among them, Branch A is located in a more developed area and the scale of supply chain is large; Branch B is located in a remote province with a smaller scale of supply chain.

4.2 Evaluation Process

Step 1. Construct fuzzy consensus decision matrix.

According to the scoring results of multiple experts, the fuzzy consistent judgment matrix of the fuzzy level evaluation of the QBENS supply chain evaluation index system is obtained. Taking the quality dimension index as an example, the fuzzy consistent judgment matrix generated by the experts after scoring is:

$$R_q = \begin{bmatrix} 0.50 & 0.51 & 0.39 & 0.69 & 0.39 & 0.69 \\ 0.49 & 0.50 & 0.39 & 0.68 & 0.38 & 0.68 \\ 0.61 & 0.62 & 0.50 & 0.78 & 0.68 & 0.78 \\ 0.31 & 0.32 & 0.22 & 0.50 & 0.22 & 0.50 \\ 0.61 & 0.62 & 0.5 & 0.78 & 0.50 & 0.78 \\ 0.31 & 0.32 & 0.22 & 0.50 & 0.22 & 0.50 \end{bmatrix} \tag{5}$$

Similarly, the fuzzy evaluation matrix of the whole index system $R_{32\times32}$ can be obtained.

Step 2. Calculate the weight of each secondary index.

According to formula (3), taking the resolution parameter $\beta = e^3$, the weights of the whole index system are calculated, as shown in Table 1.

Table 1. Weights of various dimensions and secondary indicators

Dimension	Secondary indicators							
Quality	Q1	Q2	Q3	Q4	Q5	Q6		
17.34%	2.73%	2.61%	5.07%	0.93%	5.07%	0.93%		
Benefit	B1	B2	B3	B4	B5	B6		
27.67%	10.96%	5.15%	5.15%	2.80%	2.91%	0.70%		
Efficiency	E1	E2	E3	E4	E5	E6		
13.76%	3.72%	1.05%	1.05%	5.82%	1.06%	1.06%		
Norm	N1	N2	N3	N4	N5	N6	N7	N8
27.37%	4.96%	3.51%	1.82%	3.51%	1.82%	1.82%	6.02%	3.91%
Service	S1	S2	S3	S4	S5	S6		
13.87%	6.92%	1.05%	1.05%	1.05%	1.05%	2.73%		

Step 3. Determination of comprehensive score.

According to formula (4), the scores of each dimension and the comprehensive scores of the two regions are as follows (Table 2).

Table 2. Score of each dimension and comprehensive score

	Comprehensive	Quality	Benefit	Efficiency	Norm	Service
Branch A	85.28	84.15	64.35	97.13	96.97	99.56
Branch B	83.05	71.95	98.49	67.4	77.52	92.57

4.3 Result Discussion

Based on the actual situation, the operation situation of the power grid supply chain in the two regions is analyzed through the above-mentioned evaluation index system.

Branch A is located in a developed city, and its indicators score high in the three dimensions of efficiency, norm and service, indicating that the supply chain of Branch A has efficient operation and maintenance and a high degree of standardization, providing high-quality internal and external services. Its benefit dimension score is low, reflecting the failure of Branch A to exert its scale effect. Branch B is located in a more remote area. Its index benefit score is high, and the other dimensions scores are low, indicating that its overall supply chain operation status is average. Therefore, Branch B should consider increasing capital investment, strengthening supply chain construction, and improving supply chain guarantee capabilities. On the whole, Branch B's supply chain performed well in the service dimension, but the quality management effect failed to meet expectations.

The data of the index system can be used for descriptive analytics, predictive analytics and prescriptive analytics [20]. Through vertical comparison, supply chain managers can compare historical data of indicators and evaluate the effectiveness of current management work. By setting the target value of future indicators, they can improve the ability of a certain aspect of the supply chain. Specifically, managers can evaluate the progress of the current digital transformation of the supply chain according to the secondary digital indicators, optimize management measures and better practice the green supply chain development concept according to the green indicators like green packaging category coverage.

In summary, the scores of each dimension of QBENS index system can objectively reflect the current situation and development level of power industry supply chain management in different regions to a certain extent, accurately identify the defects of the supply chain, help to further strengthen the management orientation and drive the innovative development of digital supply chain in power industry.

5 Conclusions

Starting from the development requirements of modern supply chains, this paper constructs an electric power industry supply chain management QBENS evaluation index system based on 32 indicators in 5 dimensions of quality, benefit, efficiency, norm, and service. This system can fully reflect the business of the electric power industry supply chain. The system combines the characteristics of the digital supply chain and the development concept of the green supply chain, which is forward-looking and advanced. Through the actual application of a large power grid company in China, the fuzzy analytic hierarchy process is used to set the index weights, and the branch supply chain management of different regions is evaluated and analyzed, which proves the feasibility and effectiveness of QBENS evaluation index system in the practical application of the electric power industry supply chain management.

References

1. He, M., Wang, W.: The international mirror and Chinese strategy of the development of modern supply chain. Reform (01), 22–35 (2018)
2. Fan, J., Gao, Z., Zheng, Y., Chen, D.: The construction elements, evolution and practice of digital logistics system of modern smart supply chain in power grid. Logist. Technol. **40**(02), 102–109 (2021)
3. Zhao, G., Qian, G., Wang, S.: Path analysis of green power and low-carbon development under the "dual carbon" goal. Huadian Technol. **43**(06), 11–20 (2021)

4. Fan, W., Fan, J., Li, M., Zhang, D., Luo, Y.: Construction and empirical research on the performance evaluation system of differential suppliers of electric power materials. Electr. Times **04**, 84–86 (2019)
5. Yuan, R., Liu, B., Huang, K.: Power industry supply chain risk assessment based on influence diagrams. Logist. Technol. **31**(21), 353–355+480 (2012)
6. Liang, H., Liu, Z.: Comprehensive benefit evaluation of intelligent distribution network based on AHP-entropy method-fuzzy comprehensive analysis [J/OL]. J. North China Electr. Power Univ. (Nat. Sci. Edn.), 1–8 (2021). http://kns.cnki.net/kcms/detail/13.1212.TM.20210914.1444.002.html
7. Liu, Y., Shi, H., Zeng, X.: Carbon footprint assessment of electric power enterprises based on life cycle assessment-taking a coal-fired power plant in Luliang City, Shanxi Province as an example. Resour. Sci. **33**(04), 653–658 (2011)
8. Lu, J., Yao, J., Wang, G.: Nuclear power fuel green supply chain index evaluation system. China Manag. Inf. Technol. **12**(13), 110–113 (2009)
9. Ji, Y., Yang, S.: Evaluation of power grid enterprise material management level based on combined weights. Mod. Bus. Trade Ind. **36**(22), 51–54 (2015)
10. Yang, Y., Wang, Y.: Research on the evaluation index system of power grid material supply chain. Supply Chain Manag. **1**(07), 88–94 (2020)
11. Morteza, S., Farhad, H.L., Hilda, S.: Supply chain performance evaluation with data envelopment analysis and balanced scorecard approach. Appl. Math. Model. **38**(21–22), 5092–5112 (2014)
12. Ge, W., Huang, S.H., Dismukes, J.P.: Product-driven supply chain selection using integrated multi-criteria decision-making methodology. Int. J. Prod. Econ. **91**(1), 1–15 (2004)
13. Lei, M., Ma, H., Shang, H., Ye, L.: Research on maturity evaluation of power grid material supply chain based on DDVRN. China Manag. Sci. **24**(S1), 650–657 (2016)
14. Li, Z.: Supplier performance appraisal—QCDS. China Secur. Futures **09**, 294–307 (2012)
15. Gao, X.: Research on the performance of cross-border e-commerce B2C import supply chain model. J. Anhui Normal Univ. (Humanit. Soc. Sci. Edn.) **47**(01), 134–141 (2019)
16. Mu, Y., et al.: Comprehensive evaluation index system for power grid safety and benefit based on multi-operator analytic fuzzy evaluation. Power Syst. Technol. **39**(01), 23–28 (2015)
17. Lotfi, F., Fatehi, K., Badie, N.: An analysis of key factors to mobile health adoption using fuzzy AHP. Int. J. Inf. Technol. Comput. Sci. (IJITCS) **12**(02), 1–17 (2020)
18. Liu, S., Liu, J., Xie, S., Ke, X., Chen, Y.: Analysis of the scoring method of the performance evaluation index of my country's three-level public hospitals. China Health Resour. **24**(03), 268–271 (2021)
19. Shi, J., Dong, C., Hou, J.: Research and design of teaching evaluation system based on fuzzy model. Int. J. Educ. Manag. Eng. (IJEME) **2**(10), 45–51 (2012)
20. Anitha, P., Patil, M.M.: A review on data analytics for supply chain management: a case study. Int. J. Inf. Eng. Electron. Bus. (IJIEEB) **10**(5), 30–39 (2018)

Using the Morphological Approach by the Creation of Innovative Renewable Energy Sources at the Conceptual Design Stage

Dmitry Rakov(✉)

Mechanical Engineering Research Institute of the Russian Academy of Sciences (IMASH RAN), Moscow 101000, Russia
rdl@mail.ru

Abstract. The article discussed with the creation of renewable energy systems and engineering solutions based on the morphological approach. The main disadvantage of traditional energy sources is their exhaustibility. When used, there is "thermal" pollution of the environment and the release of gases CO_2, methane, NO and others, which leads to the greenhouse effect. Non-conventional, renewable sources include hydropower, tidal energy, wave fluctuations, wind energy, solar energy, which converts solar energy into electrical and thermal energy. A special place among unconventional and renewable sources is occupied by solar and hydrogen energy. The article considers the application of the morphological approach to the study of renewable energy use. This approach is considered as part of knowledge representation and engineering. Two engineering studies are given as examples. The object under study was a transportation system using renewable energy sources. Hydrogen-based renewable energy sources were also investigated.

Keywords: Knowledge engineering · Green energy in transport · CAD design · morphological matrix · Renewable energy · Solar and hydrogen energy

1 Introduction

At present, the search for and research of alternative energy sources that allow to obtain electrical energy from renewable resources and that are environmentally friendly and economical is going on intensively. One of the main sources is the use of solar energy. Experts suggest that solar energy could provide up to 25% of humanity's electricity needs by 2050.

The sun as a source of energy has potentially almost unlimited possibilities. So, the amount of solar energy falling on the surface of the Earth is estimated at $1.5 * 10^{18}$ kW, i.e., more than 10,000 times the energy needs of mankind.

The main obstacles to the widespread use of solar energy systems remain their relatively high cost, dependence on weather conditions and low efficiency of energy conversion.

The development of ecological systems is currently one of the main challenges facing science. The catastrophic consequences of the extensive development of

Z. Hu et al. (Eds.): AIPE 2021, LNDECT 119, pp. 160–169, 2022.
https://doi.org/10.1007/978-3-030-97064-2_16

transport and its decisive impact on the environment make it necessary to search for new technical solutions allowing for a qualitative breakthrough in technology.

The modern period of science and technology development is characterized by complication of created technical means, sharp increase in the cost of their development, production, operation, as well as by rapid obsolescence and decrease in the possibility of simultaneous existence of several products of close purpose. In this connection, the task of creating equipment that embodies the latest scientific and technological achievements and discoveries and has a high engineering level is of particular importance. Research in this area refers to management and knowledge engineering [1]. Their use allows a dramatic increase in the quality of research conducted [2, 3]. In addition to research, their use in teaching is promising [4].

Studies of new engineering solutions (ES) ES should be conducted from the standpoint of system analysis.

2 Morphological Approach

The main law of development of ES is scientific and technological progress. Accordingly, the implementation of the results of scientific and technological progress is a determining condition for the creation of new equipment. All the time there is a constant updating of ES. New needs arise, which leads to the emergence of new requirements for the ES and the creation of more perfect systems. This process is described by S-curves (Fig. 1) [5].

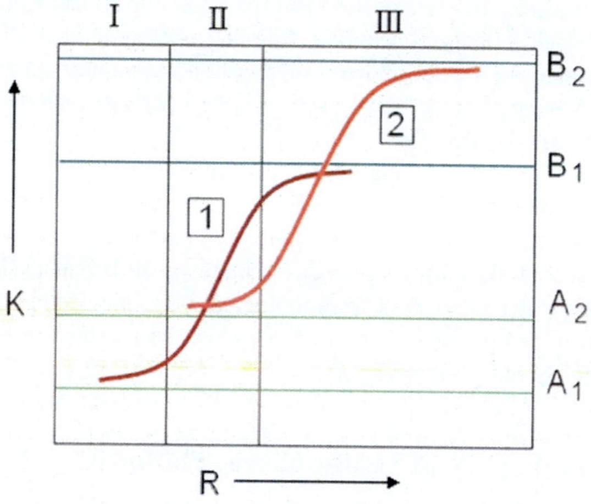

Fig. 1. Evolution of ES quality: (K - quality of ES, R - resource costs, A - socio-economic conditions of ES appearance, B - limiting values of ES quality)

At I stage (curve 1) the costs for improvement of quality of ES prevail (experimental debugging and trial operation of the new ES). At the stage II there is an intensive development of the ES with the predominant growth of quality. The effectiveness of the

ES with the selected structure and principle of action approaches the limit value - line B. If there are no ESs with a new principle of action, the system evolves for a long time. As a rule, a new principle of action of the ES appears, which in the future can provide a fundamentally higher level of quality (curve 2). The development of the new ES begins at the moment of the appearance of the need (social and economic conditions) for a new system. In development, the new ES (curve 2) initially lags behind the preceding ES (curve 1) in terms of quality, but gradually supersedes the competitor.

To make progress, it is necessary to use CAD in the conceptual design phase.

However, so far, the high expectations placed on the use of computational tools in the design process are not fully justified. In practice, the design of a device (system, process) should be considered as a set of two main tasks: the choice of structure, or structural synthesis, and the choice of numerical values of parameters of the elements of this structure, or parameter synthesis.

In these terms we can consider the design of almost any object at any stage of its creation. The methods for solving these problems differ significantly. Their complexity also differs. Tasks of parametric synthesis, as a rule, are reduced to the search of solutions satisfying the metric criteria, which makes them formally solvable. The problem of structural synthesis is quite different and cannot be generally referred to the class of formally solvable problems. The result of structural synthesis is the choice of a rational structure of the object. Synthesis of the structure, with varying degrees of detail can be carried out at various stages of the design process [6].

However, the greatest effect can be obtained at an early stage of the design process - the technical proposal stage. A characteristic feature of the technical proposal stage is the limited information about the properties of the future system, which makes, first of all, to refer to the structure of the system and the information contained in it. Methods of morphological analysis deal with this successfully.

Morphological methods have a number of disadvantages, the main of which are the labor-intensiveness of selecting from a variety of alternative ES variants, the complexity of assessing the effectiveness of a particular generated option, and in some cases the impossibility of analyzing all possible variants [7, 8].

To eliminate the above disadvantages, the methodology and software package "Okkam" [8, 9], based on the provisions of system, cluster analysis and morphological approach, were developed. Advanced method of ES analysis and synthesis is a development of F. Zwicky's "morphological box" method. Fritz Zwicky formulated: "The goal of morphological research is to see the perspective of a complete "field of knowledge" about a subject" [7].

3 Power Supply of Stratospheric Platforms

For the analysis and synthesis was selected class of stratospheric platforms of long-term basing. Promising transport systems are so-called High Altitude Platform Station HAPS), based at altitudes of 19–22 km [10]. The architecture of all stratospheric systems is largely similar: the aircraft plays the role of the top of a wireless broadband network serving a highly urbanized region. The system provides direct transmission of data between users within the area served, while public networks are used to communicate with external recipients. The creators of the systems talk about "ubiquitous access" for their customers, implying that there are no "obstacles" in the form of vegetation, buildings or physical features of the terrain between the high-altitude platform and the subscriber (Fig. 2). For Russia very promising is the use of such systems to supply autonomous residential complexes.

All stratospheric platforms have virtually the same architecture: the aircraft plays the role of the top of a wireless broadband network serving a given area. Mobile users will receive services of digital telephony, fax messages and the Internet at 64 kbit/s (stationary users will be able to transmit information at speeds up to 2 Mbit/s.) At each moment of time one high-rise platform can provide up to 400 thousand users. This means that one stratospheric platform above Moscow could serve up to 14 million cell phone subscribers.

The main tasks for high-altitude platforms are:

- research (atmospheric monitoring and research);
- communications;
- observation area.

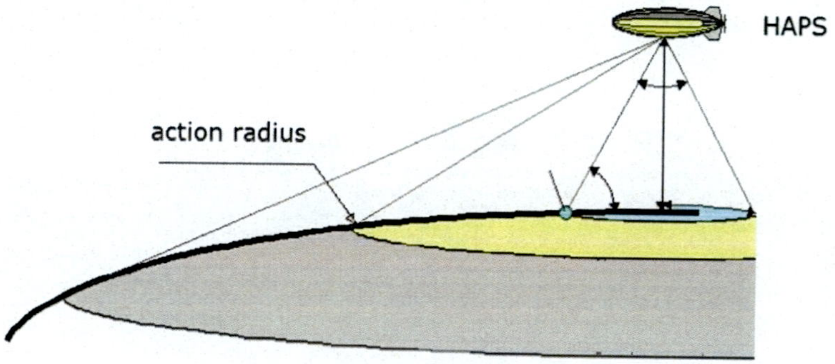

Fig. 2. Scheme of the stratospheric platform action

The configuration, composition and program of operation of the vehicles that meet the same requirements may differ significantly from each other. One of the most important problems in design is the choice of energy storage systems to ensure long-term cruising of vehicles in the stratosphere.

Building a morphological table and exploring variants Variants of different configurations of new stratospheric systems have been studied in detail [11]. As a result of decomposition and analysis the morphological matrix (MM) is constructed (Fig. 3). Stratospheric aircraft and balloon projects were selected as reference variants [12]. Based on the results of the analysis, a classification field of possible technical solutions was built.

	P^1	P^2	P^3	P^4
P_1	P_1^1	P_1^2	P_1^3	
P_2	P_2^1	P_2^2	P_2^3	
P_3	P_3^1	P_3^2		
P_4	P_4^1	P_4^2	P_4^3	
P_5	P_5^1	P_5^2	P_5^3	
P_6	P_6^1	P_6^2	P_6^3	
P_7	P_7^1	P_7^2	P_7^3	P_7^4
P_8	P_8^1	P_8^2		

Fig. 3. Morphological matrix of aerostats (screenshot)

The power of the morphological set is equal to 5180 variants. 123 variants located in 7 clusters were selected for analysis (Fig. 4). As a result, several promising systems were synthesized (Fig. 5).

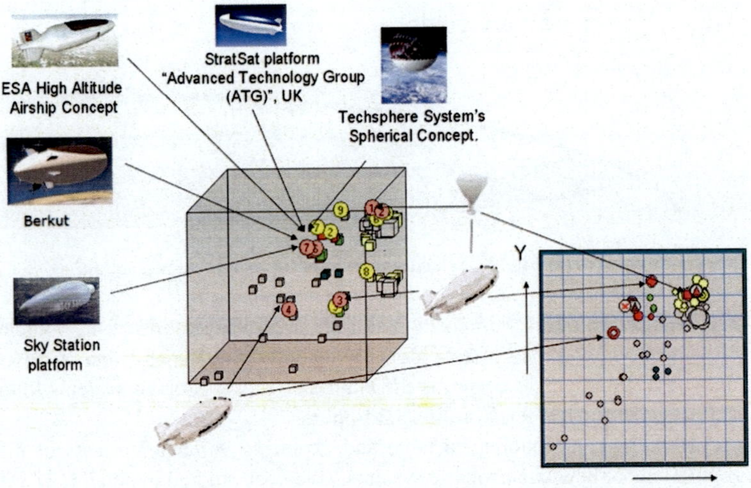

Fig. 4. Clustering and comparison of stratospheric platforms by different energy supply systems

	P^1	P^2	P^3	P^4
P_1	**P_1^1**	P_1^2	P_1^3	
P_2	P_2^1	**P_2^2**	P_2^3	
P_3	**P_3^1**	P_3^2		
P_4	P_4^1	**P_4^2**	P_4^3	
P_5	P_5^1	**P_5^2**	P_5^3	
P_6	P_6^1	**P_6^2**	P_6^3	
P_7	**P_7^1**	P_7^2	P_7^3	P_7^4
P_8	P_8^1	**P_8^2**		

Fig. 5. The rational engineering solution for a synthesized aerostat

4 Parametric Optimization and Variants Comparison

The critical technology in the creation of stratospheric unmanned aerial vehicles and free balloons is the provision of electricity, and accordingly, the choice of suitable energy sources. During the day, solar cells provide energy for the engines that keep the vehicles in a given place, as well as communication equipment.

To work at night, electricity can be stored in rechargeable batteries, but they have a lot of weight, so now the most promising is the use of fuel cells. Principle of operation of onboard power supply system looks as follows (Figs. 6 and 7).

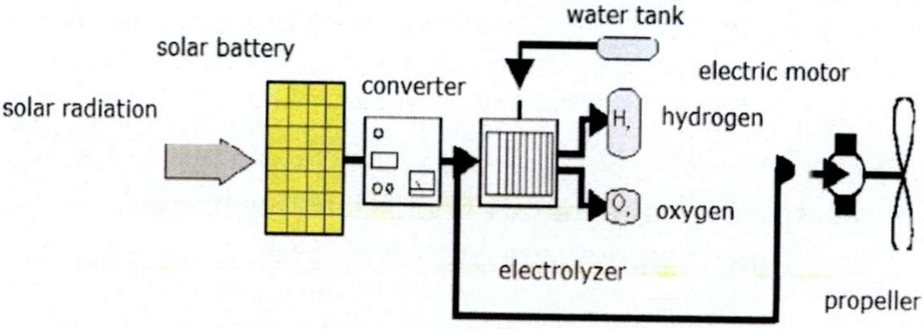

Fig. 6. Power supply of airship systems in the daytime

During daytime solar batteries directly supply propulsion system and equipment, and part of energy is used for splitting of water (there is a water tank onboard the aerostats) into hydrogen and oxygen in the electrolyzer.

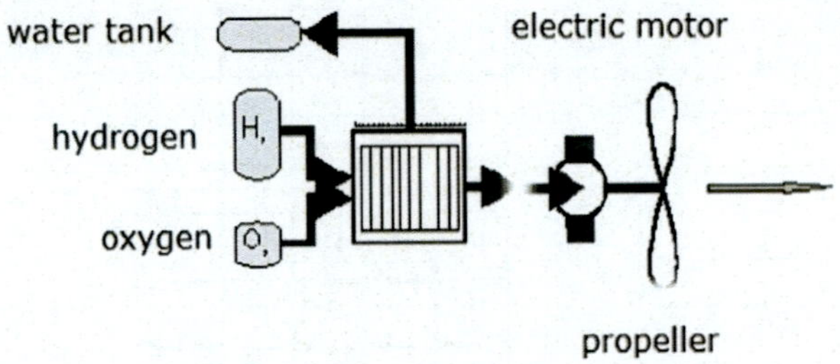

Fig. 7. Powering the airship systems at night

At night, the stored hydrogen and oxygen are converted into water in the fuel cells to form electrical energy, and the resulting water is stored in the tank. On the next day the process is repeated. A promising method of producing hydrogen from hydride batteries, which consist of hydride cartridges filled with metal hydride powder, is considered promising. The powder is based on lanthanum nickel hydride. When it is heated, hydrogen is released, and as the pressure increases, the molecules of the material intensely absorb hydrogen (each alloy molecule holds six hydrogen atoms). By placing the unit on board, the balloon, the autonomy of the flight can be significantly increased. Hydrogen is used both to feed the envelope and to generate electricity, as well as to fuel the aerostat's engines. Such balloons could become the most environmentally friendly vehicles. The use of hydrogen as a fuel is also convenient because its combustion does not change the mass of the aerostat. 1 kg of hydrogen obtained from 150 g of metal-hydride powder has the same heat of combustion as 2.5 L of kerosene. It is also possible to create a helium-hydrogen balloon with hydrogen compartments inside the helium shell.

Another possibility for powering unmanned aerial vehicles or free balloons is to use microwave radiation transmitted to the vehicle from Earth. This technology has been tested in Japan with 30 kW of power transmitted over a distance with an energy transfer efficiency of 54%. The disadvantages of this technology, first of all, are the need to create exclusion zones and disastrous environmental friendliness.

5 Morphological Analysis Hydrogen Generation for Renewable Energy Systems

Morphological analysis and experimental work on hydrogen generation systems were performed for the systems under study.

Hydrogen is environmentally safe because it turns into ordinary water vapor when burned. The main problem is storage of hydrogen. A known method is storage in chemical compounds in which hydrogen is contained in a bound form. For storage systems can serve: compounds CH_4, C_2H_6, C_3H_8, C_4H_1 and water with energy storage materials from oxides in which the hydrogen content can be 10% by mass [13].

The technological scheme consists of three stages:

- Obtaining energy storage materials using a primary energy source.
- Obtaining hydrogen (and possibly thermal energy) using energy storage materials.
- Use of hydrogen as a fuel.

To analyze possible alternatives, a morphological analysis is used and a morphological matrix is formed.

Table 1. Morphological matrix

Feature	Option 1	Option 2	Option 3	Option 4
Energy sources	Non-renewable	Wind	Solar energy	Nuclear power
Hydrogen production	Hydrogen from biomass	Steam conversion	Electrolysis	Gasification coal
Hydrogen storage systems	Liquid H	Gaseous H	Generation of hydrogen by reaction A1 with Mg and Cu	Absorption of H in the Al melt
Receiving energy	Energy fuel cells	Combustion H		
Target use	Electricity generation	Road transport	Air transport	Water transport

After the synthesis and analysis stages, several solutions remained (Fig. 8). The advantages of the generated solutions are environmental friendliness, safety, mass and cost performance. A special feature of the synthesized solutions is the use of hydrogen storage alloys for transport aerostatic systems.

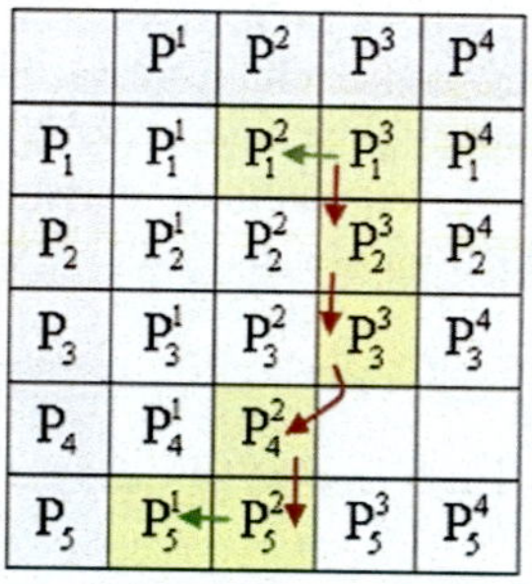

Fig. 8. Selected solutions in the morphological matrix

6 Results and Discussion

A1-based alloys, excluding the corrosion-inhibiting elements Mn, Ti, Ni and Si, were made for the experiments (Table 2) [13]. The results of gas emission measurements are shown in Fig. 9.

Table 2. Investigated alloys

Weight composition, %					
Alloy	Al	Cu	Mg	Zn	Fe
1	96	4	–	–	–
2	99	1	–	–	–

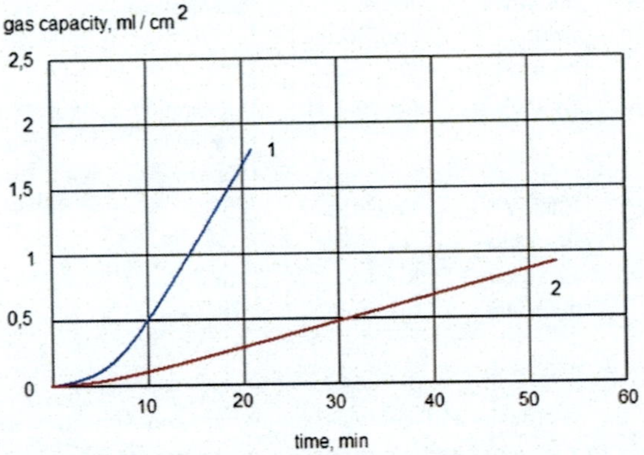

Fig. 9. Hydrogen gas capacity for aluminum with 4% copper (line 1) and for aluminum alloy with 1% copper (line 2)

The most rational is an Al alloy with 4% copper. Activated aluminum serves as a fuel and releases hydrogen when reacting with water. The reaction products are completely harmless and form a solid insoluble powdery residue [14, 15].

The considered morphological approach can be considered as a significant alternative to the methods of multi-attribute decision making [16]. In the future it is expected to give a comparison of these methods with an indication of their advantages and disadvantages.

7 Summary and Conclusion

The article deals with the problems of synthesis renewable energy transport systems and engineering solutions based on the morphological approach. The approach and software allow generating a morphological set of variants at the stage of conceptual design.

Among the unconventional and renewable sources, solar and hydrogen energy systems were investigated. Two engineering studies are given as examples. The object under study was a transportation system using renewable energy sources. Hydrogen-based renewable energy sources used for the transportation system were also investigated. Experimental studies have also been conducted on the use of hydrogen as an energy carrier.

The use of an integrated methodology allows new innovative technologies and solutions to be identified. The proposed approach is considered as part of knowledge engineering.

References

1. Filemon, A., Uriarte, J.: Introduction to Knowledge Management. ASEAN Foundation, Jakarta (2008)
2. Zaied, A.N.H., Hussein, G.S., Hassan, M.M.: Role of knowledge management in enhancing organizational performance. IJIEEB **4**(5), 27–35 (2012)
3. Keishing, V., Renukadevi, S.: A review of knowledge management based career exploration system in engineering education. Int. J. Mod. Educ. Comput. Sci. (IJMECS) **8**(1), 8–15 (2016)
4. Njenga, S., Oboko, R., Omwenga, E., Maina, E.: Use of intelligent agents in collaborative m-learning: case of facilitating group learner interactions. Int. J. Mod. Educ. Comput. Sci. **10**, 18–28 (2017)
5. Silin, V.B.: Search for Structural Solutions by Combinatorial Methods. MAI, Moscow (1992)
6. Rakov, D.: Okkam - advanced morphological approach as method for computer aided innovation (CAI). In: MATEC Web of Conferences, ICMTMTE 2019, vol. 298 (2019)
7. Zwicky, F.: Discovery, Invention, Research - Through the Morphological Approach. The Macmillan Company, Toronto (1969)
8. Ritchey, T.: General morphological analysis as a basic scientific modelling method. Technol. Forecast. Soc. Chang. **126**, 81–91 (2018)
9. Bardenhagen, A., Rakov, D.: Advanced morphological approach in aerospace design during conceptual stage. Facta Univ. Ser.: Mech. Eng. **17**(3), 321–332 (2019)
10. Kröplin, B.: High altitude plattforms. In: 3rd Airship Convention and Exhibition in Friedrichhafen, pp. 6–12 (2000)
11. Rakov, D., Thorbeck, J., Böhm, F.: Vergleich von verschiedenen Konzepten für aerostatische stratosphärischen Plattformen, pp. 1217–1224. DGLR-Kongress, Stuttgart (2002)
12. Toyer, C., Grace, D.: High altitude platforms for wireless communications. Electron. Commun. Eng. J. **13**, 127–137 (2001)
13. Tereshchuk, V., Rakov, D.: The technology of hydrogen generation based on active metals. In: International Multi-Conference on Industrial Engineering and Modern Technologies (FarEastCon), Vladivostok, pp. 1–5 (2020)
14. Esman, V., Tereshchuk, V., Rakov, D., Kozlyakov, V., Pecheikina, M.: Analysis of systems using renewable energy sources and materials for hydrogen generating. Heavy Eng. **10**, 11–13 (2010)
15. Tereshchuk, V.: Application of Energy Active Metals and Hydrogen. Innovative Mechanical Engineering, Moscow (2019)
16. Adriyendi, M.: Multi-attribute decision making using simple additive weighting and weighted product in food choice. Int. J. Inf. Eng. Electron. Bus. (IJIEEB) **7**(6), 8–14 (2015). https://doi.org/10.5815/ijieeb.2015.06.02

Automation of the Monitoring in Metal Cutting Operations as Fast-Variable Processes Using Artificial Intelligence Methods

Sergei Dosko, Vladimir Utencov, Aleksey Spasenov(✉), Igor Lukashin, and Kirill Kucherov

Bauman Moscow State Technical University, Moscow 105005, Russia

Abstract. The paper considers the process of creating an automated monitoring system for the cutting process. The proposed system, based on the analysis of fast-variable processes in real time, will allow evaluating the machine data, notifying decision-makers at the time of anomalies in the technological process, and indicates the possible cause of the anomaly and gives recommendations on further actions. The system using artificial intelligence methods makes it possible to determine in advance the wear and tear of the cutting tool of the machine tool, operator errors, machine malfunction, increased hardness of the workpiece and the presence of foreign inclusions in the workpiece, as well as incorrect fastening of the part and displacement of the part during processing. The experience of the development and use of machine learning methods for the analysis of vibroacoustic signals in the monitoring system of the processing process on machine tools is presented. The results of using the approximation and structural parametric approach for identifying cutting conditions are shown. As a result of the research, the effectiveness of the use of machine learning methods at each stage of experimental data processing has been confirmed.

Keywords: Time-series data · Automated technological environment · Spectral analysis · Approximation approach · Structural parametric analysis

1 Instruction

Currently, the industry maintains a tendency to increase productivity and processing accuracy [1–5]. An increase in the productivity of the machine can be achieved by increasing the cutting speed, however, at high cutting speeds, vibrations inevitably increase, which negatively affect the surface roughness [6] and tool life [7]. Therefore, it is necessary to control the vibration level in the machine within the permissible limits and avoid exceeding it. For this, it is necessary to develop a system for monitoring the state of the machine during the cutting process.

An increase in vibration in the machine above the permissible level can serve as a criterion for an unstable cutting mode. However, for each machine, tool and material to be processed, the allowable vibration level will be different. Therefore, this approach is ineffective and poorly subject to automation. The article proposes a different principle for determining the unstable cutting mode, which is invariant with respect to various types of machines, tools and workpiece material and is well subject to automation.

Z. Hu et al. (Eds.): AIPE 2021, LNDECT 119, pp. 170–180, 2022.
https://doi.org/10.1007/978-3-030-97064-2_17

The use of systems and the accumulation of experimental data based on artificial intelligence methods in the processing of products makes it possible to improve the methods of identifying the wear of a working tool, for example, by vibroacoustic signals. Such systems make it possible to use the method of self-tuning of diagnostic models of wear of working tools based on the data obtained during the operation of the system. The creation of algorithms for predicting breakdowns should be based on a priori data that consider the nonlinearity of wear functions, thereby defining models that describe the operation of systems [8–11].

To automate vibration detection, machine learning methods are used that perfectly cope with the tasks that arise at each stage of experimental data processing (preprocessing, extraction of characteristic features, selection of the most informative and their classification).

The purpose of this work was: research of machine learning methods for automatic determination of the state of the cutting process by sound and vibration signals.

2 Monitoring Automation Stages

In the process of creating intelligent monitoring systems, problems arise associated with the formation of data, the choice of characteristic features and models that describe the process under study. Initial data should contain all stages of material processing, and characteristics and models must be selected in such a way that they allow interpretation of the results.

2.1 Methodology

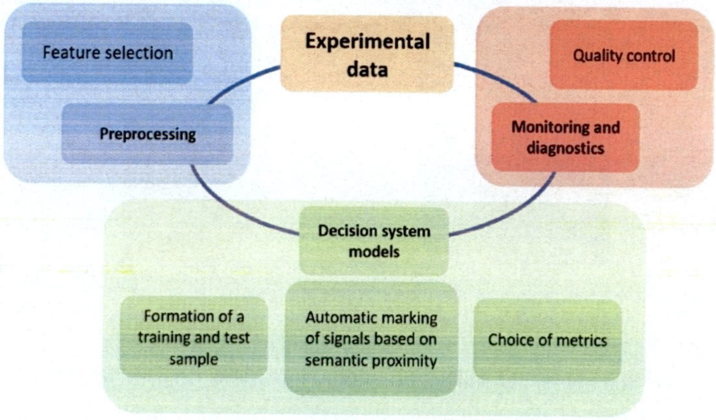

Fig. 1. Data processing and decision systems

The development of means for recording signals makes it possible to analyze the operation of machine tools in various frequency ranges. The widespread use of appropriate measuring systems has influenced the exponential growth of data that needs to be processed qualitatively. To create expert systems, it is required to use markup assessors, which is a very resource-intensive task. To solve this problem and to create effective monitoring systems, it is proposed to use the scheme proposed in the Fig. 1. This approach uses approximation approaches to analyze local transient processes with the subsequent formation of decision rules to identify each stage of the cutting process.

2.2 Description of Experimental Data

Data collection is carried out during an experiment to receive vibration signals from an accelerometer and a sound signal from a microphone under various operating modes of the machine: idling, with stable and unstable operating modes [12–14]. To obtain experimental data, a Tongtai CT-350 5-axis machining center was used. Workpiece material - Al6061-T6. The diameter of the end mill is 12 mm, the angle of inclination of the helical groove is 26°, the number of cutting edges is 2. In the experiment, the accelerometer was mounted on the body of the spindle assembly above the supports. A microphone was attached near the spindle. In Fig. 2 shows an experimental data collection setup.

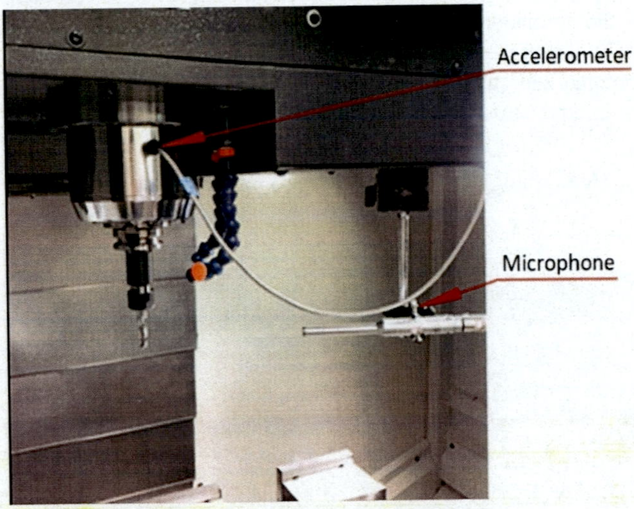

Fig. 2. Experimental setup

Table 1 shows the values of cutting conditions used in the experiment.

Table 1. Cutting conditions

Measurement number	1	2	3	4	5	6	7	8	9	10
Feed rate, mm/min	150									
Spindle speed, rpm	3000						3500	3000	3500	3000
Cutting depth, mm	–	–	0.6	1.0	1.2	1.4	0.5	1.0	1.5	1.5
Working hours	Idling		Stable		Unstable		Idling		Stable	

2.3 Spectral Analysis of Signals Based on the Approximation Approach in the Time Domain

The main requirements for the method of spectral analysis of a signal - smoothing the spectral picture and increasing the resolution - are contradictory. The known methods of dealing with the shortcomings in the framework of the classical Fourier transform (FT) are to use weight windows, but this leads not only to smoothing the spectral picture, but also to deterioration in the resolution. In Fig. 3 and 4 show the original vibration signal and its spectrogram, respectively.

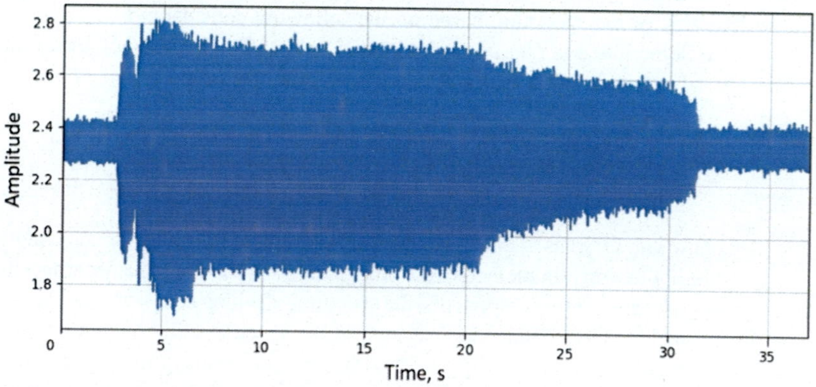

Fig. 3. Vibration signal

One of the ways to overcome this contradiction, according to Volkov N.V. [15] and Myasnikova N.V. [16, 17], is to use an approximation approach. In the tasks of monitoring the cutting process, the form of the approximating function must be related to the target task and must also be universal enough to fit a wide class of signals. Most technical problems require approximation by functions reflecting physical processes. In this case, the mathematical model should be determined by a finite set of informative parameters and adequately describe the process under study within the framework of the task [16].

When analyzing the dynamics of the process under study, a qualitative picture is usually more important and informative than a quantitative one, and a complete qualitative understanding of the process is carried by its extreme values. In [16], the following

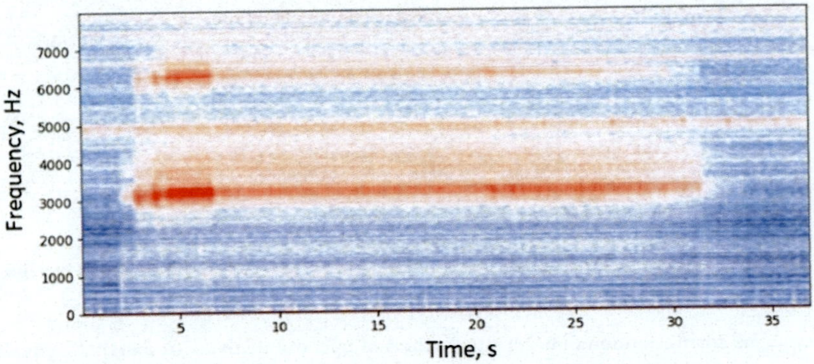

Fig. 4. Spectrogram

limiting version of the Kotelnikov theorem is proposed: any multiextremal function with a bounded spectrum can be reconstructed with an accuracy sufficient for practical purposes from the extreme values of the function and the intervals between them.

The linearity of the FS allows us to apply a general approach to the approximate calculation of the spectrum, based on the following considerations.

Let some function $f(t)$ be approximated with sufficient accuracy by a finite sum of arbitrarily chosen functions $f_k(t)$, i.e.

$$f(t) \approx \sum_{k=1}^{N} f_k(t)$$

This ratio can serve as the basis for various variants of formulas for calculating spectra.

The approximation method of spectral analysis is based on the decomposition of the signal under study into components, the spectrum of which is known, and the spectrum of the original signal is obtained as the sum of the spectra of all components. Mathematically, the method is based on approximating the signal in the time domain with any given accuracy by basic functions of the form e^{-x^2}, $ch^{-1}(x)$, $1/(1+x^2)$, i.e. "Bell" impulses.

With this approach, the real signal is replaced by functions of the form

$$f(t) = \sum_{i=1}^{l} \sum_{k=1}^{m_i} \alpha_{ki} \varphi[\beta_{ki}(t - c_{ki})],$$

and the spectrum is determined by the expression

$$S(j\omega) = \sum_{i=1}^{l} \sum_{k=1}^{m_i} \frac{\alpha_{ki}}{\beta_{ki}} \varphi\left(\frac{\omega}{\beta_{ki}}\right) e^{-j\omega c_{ki}},$$

where $\alpha_{ki}, \beta_{ki}, c_{ki}$ are the parameters of the approximating function; m is the number of extrema: m_l - in the original implementation of the signal, $m_1, m_2, \ldots, m_i$ - in the difference between the original implementation and the approximating function; l is the number of iterations required to achieve a given precision.

The authors of [13] comprehensively investigated the question of the convergence of the approximation procedure by Gaussian functions of the form $e^{\beta^2 x^2}$

$$f(t) = \sum_{i=1}^{l} \sum_{k=1}^{m_i} x_{ki} e^{\beta_{ki}^2 (t - c_{ki})^2 \beta}$$

and the expression for the complex spectrum has the form

$$S(j\omega) = \sqrt{\pi} \sum_{i=1}^{l} \sum_{k=1}^{m_i} \frac{x_{ki}}{\beta_{ki}} e^{-\omega^2 / \left(4\beta_{ki}^2\right) - j\omega c_{ki}}$$

where: x_{ki} - extrema: x_{kl} for the original signal, presented as a series, $x_1, x_2, \ldots, x_{ki}$ - for the difference between the implementation and the approximating function.

It is shown that the procedure of approximation by extreme signal values is the most effective for monitoring fast processes.

2.4 Identifying Cutting Conditions Using Structural Parametric Analysis

It is shown that the procedure of approximation by extreme signal values is the most effective for monitoring fast-variable processes.

A feature of the analysis of complex technical systems (CTS) is the simultaneous presence of various processes in the segments of the original time series. Each cutting mode corresponds to a characteristic process represented in the signal. Within the framework of this assumption, the task of analyzing the initial signal is the task of determining the fact of the transition between the states of the investigated CTS (between modes). It can be solved only if the system possesses the fundamental property of observability, i.e. from the information at the output, you can completely restore information about the states of the system. Using the idea of an approximation approach and the method of structural parametric analysis [18, 19], it is possible to solve the problem of identifying the processes occurring in the system during material processing for each cutting mode. Each process will correspond to a certain state of the CTS.

We will use Prony's method [20] to compare the observed data X, which have N complex samples $x_1, x_2, \ldots, x_N$, with the sum of complex functions:

$$Y = \sum_{k=1}^{M} A_k e^{(n-1)(\alpha_k + j2\pi f_k)\Delta + j\theta_k}$$

for $n = 1, 2, \ldots, N$, where $j^2 = -1$, and Δ is the sampling interval. The objects of estimation are the amplitude of the complex exponents A_k, the attenuation coefficient α_k, the harmonic frequency f_k and the phase θ_k. As a result of such a comparison, each segment of the time series will be described by a set of M modal parameters.

Further, using the method of soft clustering, when each set of modal parameters can relate to several typical processes, it becomes possible to compare all segments of the time series with ongoing processes, the duration of which can be significantly longer than the duration of the segments themselves. The problem of soft clustering can be considered as thematic modeling [21] over data, where each segment of the time series is a document, and a word in a document is a set of modal parameters. This approach has a significant margin of flexibility to handle complex data. According to the formula of total probability and the hypothesis of conditional independence, the distribution of sets of modal parameters in a segment is described by a probabilistic mixture of distributions of sets of modal parameters in the processes:

$$p(w|d) = \sum_{t \in T} p(w|t,d)p(t|d)$$

where w is a set of modal parameters, d is a segment associated with process t.

In Fig. 5 shows the result of identification of typical processes occurring in the system.

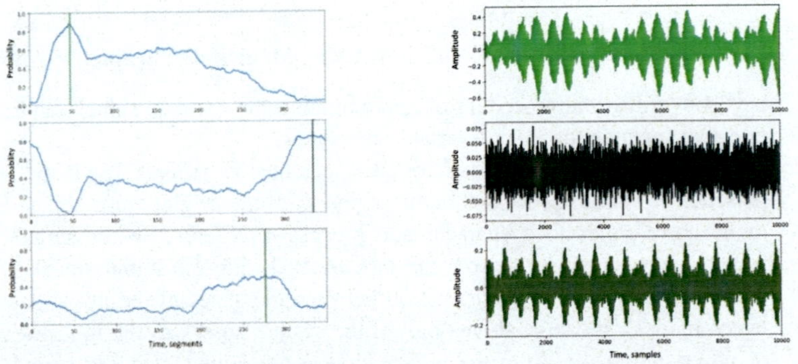

(a) Probability of segment belonging to processes

(b) Typical processes

Fig. 5. Process identification

2.5 Cutting Data Classification

For real-time monitoring, signal processing using Fast Fourier Transform (FFT) can be used. In Fig. 6 shows the spectra of vibration signals at idle (a), at stable (b) and unstable (c) modes of the cutting process.

An analysis of the results of conversions under various operating modes showed that at idle all vibration energy is concentrated in the low frequency region near the spindle speed (50 Hz) and its harmonics, the vibration amplitude in this case is an order of magnitude less than during cutting. With a stable cutting process, the main energy is concentrated around the resonant frequency of the cutting process (3200 Hz). During an unstable cutting process, the vibration energy is redistributed to the high frequency region.

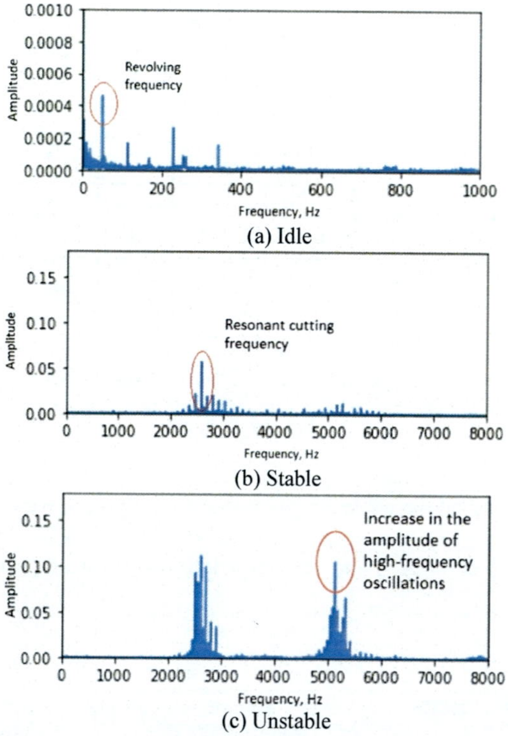

Fig. 6. Spectra of vibration signals at various modes of the cutting process

To automate the detection of the nature of vibration, the entire frequency range is divided into three characteristic regions: 0–1000 Hz; 1000–5000 Hz; 5000–8000 Hz, then the total energy is calculated in each area. Thus, each mode of operation will have its own characteristic energy distribution over the selected frequency regions.

In the next step, a learning model was created based on several classification methods to compare their accuracy. The following classification methods were used based on support vector machine (SVM), k-nearest neighbors (kNN), artificial neural network (ANN) methods. To train the models, it is necessary to form an array of training and test data, in a ratio of approximately 70% to 30%, where 70% is data for training, and 30% is data for testing. To form a data array, each of the 10 experimental signals is divided into 10 parts to increase the amount of input data. Then FFT of all the sliced signals is performed and characteristic frequency ranges are allocated in them. Next, the total energy is calculated for each frequency range. After that, the obtained data are entered into a table for subsequent training and testing of models.

The method based on artificial neural networks showed the best recognition accuracy (more than 93%), in Fig. 7 shows the matrix of errors. Mark "1" corresponds to idle speed; mark "2" corresponds to a stable cutting process, mark "3" to an unstable cutting process.

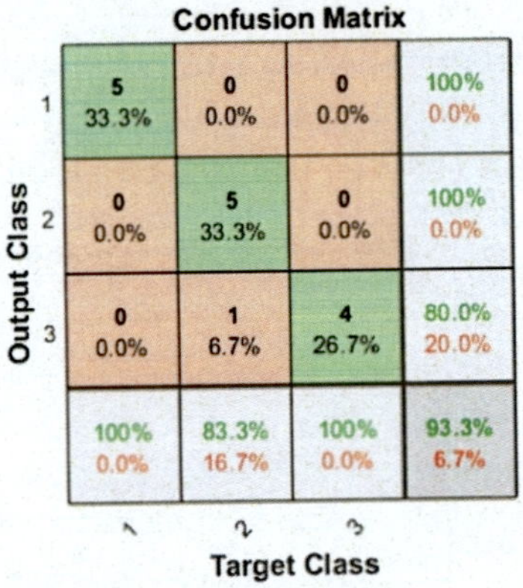

Fig. 7. Confusion matrix

Combining all the above stages into a single system, we will form a block diagram of the cutting process monitoring system (Fig. 8). Each block of the proposed system can be easily automated and combined into one software package.

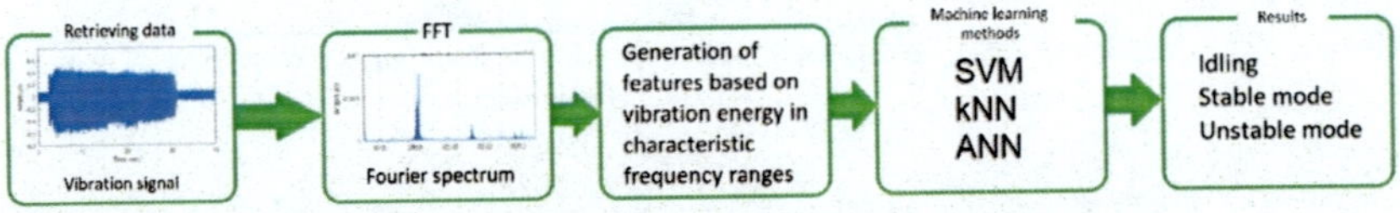

Fig. 8. Block diagram of the cutting process monitoring system

3 Discussion

The use of modern systems and methods of data analysis makes it possible to extract complex patterns in a large amount of data. Approximation approaches are well suited for handling fast-variable processes for feature extraction problems. The combination of supervised and unsupervised teaching methods improves the quality of decision support systems for monitoring the cutting process. Further work in this direction involves the creation of experimental databases containing semantically similar processes for the analysis of structurally similar objects.

4 Summary and Conclusion

The problem of detecting characteristic processes in the system and exceeding the vibration level can only be solved by introducing a machine tool monitoring system, which would include data collection, processing of received signals and vibration detection using artificial intelligence methods. In this work, a phased structure of a machine tool monitoring system was built, which includes data collection using a microphone and an accelerometer, processing the received signals using structural parametric analysis and FFT methods, as well as vibration detection using machine learning models. It is shown that the classification accuracy can be more than 93%.

Acknowledgment. This project is supported by National Taiwan University of Science and Technology (Taiwan TECH), Chung-Hsien Kuo, Ph.D., Professor and Chairman, Department of Electrical Engineering, Intelligent Vision and Autonomous Mobility Laboratory (iVAM Lab.), Dr. Minh-Quang Tran, postdoctoral, Taiwan.

References

1. Kachan, A.Y., Mozgovoy, A.Y., Belikov, S.B., Vnukov, Y.N., Karas, V.P.: Main directions of development of progressive technologies and CNC metal-cutting machines. Bull. dvigunobuduvannya, **2**, 82–85 (2007)
2. Salem, M., Khelfi, M.F.: Sequential adaptive RBF-fuzzy variable structure control applied to robotics systems. Int. J. Intell. Syst. Appl. (IJISA) **6**(9), 19–29 (2014). https://doi.org/10.5815/ijisa.2014.09.03
3. Upadhyay, V., Pandey, G.N.: Robot based preventive maintenance system for in-service inspection of equipments. Int. J. Intell. Syst. Appl. (IJISA) **7**(4), 47–53 (2015). https://doi.org/10.5815/ijisa.2015.04.07
4. Rafsunjani, S., Safa, R.S., Al Imran, A., Rahim, S., Nandi, D.: An empirical comparison of missing value imputation techniques on aps failure prediction. Int. J. Inf. Technol. Comput. Sci. (IJITCS) **11**(2), 21–29 (2019). https://doi.org/10.5815/ijitcs.2019.02.03
5. Salmanov, K., Harb, H.: Data analysis for the aero derivative engines bleed system failure identification and prediction. Int. J. Intell. Syst. Appl. (IJISA) **13**(6), 13–24 (2021). https://doi.org/10.5815/ijisa.2021.06.02
6. Tran, M.-Q., Liu, M.-K., Tran, Q.-V.: Milling chatter detection using scalogram and deep convolutional neural network. Int. J. Adv. Manuf. Technol. **107**(3–4), 1505–1516 (2020). https://doi.org/10.1007/s00170-019-04807-7
7. Chung, C., Tran, M.-Q., Liu, M.-K.: Estimation of process damping coefficient using dynamic cutting force model. Int. J. Precis. Eng. Manuf. **21**(4), 623–632 (2020). https://doi.org/10.1007/s12541-019-00297-5
8. Huang, H., Baddour, N., Liang, M.: Bearing fault diagnosis under unknown time-varying rotational speed conditions via multiple time-frequency curve extraction. J. Sound Vib. **414**, 43–60 (2018). https://doi.org/10.1016/j.jsv.2017.11.005
9. Liu, M.K., Tran, M.Q., Chung, C., Qui, Y.W.: Hybrid model- and signal-based chatter detection in the milling process. J. Mech. Sci. Technol. **34**(1) (2020). https://doi.org/10.1007/s12206-019-1201-5
10. Kozochkin, M.P., Solis, N.V.: Investigation of the relationship between vibrations during cutting and the quality of the resulting surface. Bull. Peoples'. Friendship Univ. Russia Ser.: Eng. Res. **2**, 16–23 (2009)
11. Antsev, A.V., Chong, D.H.: Prediction of cutting tool life based on vibration control during milling. Bull. Tula State Univ. Tech. Sci. **7**, 3–11 (2008)
12. Sergienko, A.B.: Digital Signal Processing, p. 751. Piter, Moscow (2003)
13. Kuljanic, E., Sortino, M., Totis, G.: Multisensor approaches for chatter detection in milling. J Sound Vib. **312**(4–5), 672–693 (2008)

14. Singh, K., Singh, R., Kartik, V.: Comparative study of chatter detection methods for high-speed micromilling of Ti6Al4V. Prod. Manuf. **1**, 593–606 (2015)
15. Volkov, N.V.: Functional series in the problems of dynamics of automated systems, p. 100. Yanus-K (2001)
16. Myasnikova, N.V., Beresten, M.P.: Express Analysis of Signals in Technical Systems, p. 180. Penza, PSU (2012)
17. Myasnikova, N.V., Beresten, M.P.: Theoretical foundations of express analysis. In: Proceedings of Higher Educational Institutions Volga Region. Technical Science, no. 6, pp. 117–123 (2006)
18. Zhuk, D.M., Volosatova, T.M., Spaseonov, A., Kucherov, K.V.: Evaluation of dynamic systems using modal linguistic analysis of multivariate time series. Dyn. Complex Syst. **14** (1), 38–45 (2020)
19. Spaseonov, A.Yu., Kucherov, K.V., Volosatova, T.M., Zhuk, D.M.: Assessment of the state of complex technical objects using structural-modal analysis of quasiperiodic time series. Inf. Technol. **26**(10), 563–569 (2020)
20. Marple, S.L.: Digital spectral analysis with application, p. 256 (1986)
21. Vorontsov, K.V., Potapenko, A.A.: Regularization, robustness and sparsity of probabilistic topic models. Comput. Res. Model. **4**(4), 693–706 (2012)

Virtual Simulation of the Surgery of Installing Transitional Implant Dentures for the Two-Stage Dental Implant Osteointegration Period

Tatiana Poliakova[1,2(✉)], Sergei Gavriushin[1], and Sergey Arutyunov[3]

[1] Bauman Moscow State Technical University, Moscow 105005, Russia
tpolyakova@bmstu.ru

[2] Space Research Institute of the Russian Academy of Sciences, Moscow 117997, Russia

[3] A.I. Evdokimov Moscow State University of Medicine and Dentistry, Moscow 127473, Russia

Abstract. In work the deep biomechanical analysis assuming creation of model in a finite-element complex is described. Computer simulation was carried out in Mimics, SolidWorks, Patran/Nastran, Ansys and Geomagic systems. Considered that if the transitional implant can't be positioned in one row with two-stage implant, then the arrangement on diagonal between them is possible. We discuss two approaches to simulation and optimization of transitional implants arrangement: optimization is considered on a simpler model with the previously set geometry, total calculation is carried out on the full-scale model constructed according to the tomogram. Options of accounting the density of the bone tissue according to classification by Misch and also model of strength characteristics distribution in a bone on the basis of density assessment to the tomogram are described.

Keywords: Transitional implants · Prosthodontics · Finite-element method · Strength · Dentistry

1 Instruction

The questions connected with rational design and use of dental implants for the last years remain relevant and take attention of specialists in biomechanics [1, 2]. The modern biomechanics is based on complex use of opportunities of medical and scientific-technical disciplines, such as theoretical mechanics, resistance of materials, theory of elasticity, numerical methods (in particular, a finite-element method) and so on, with application of a computer tomography not only for diagnostics, but also for three-dimensional simulation with attraction of data 3D-scans with the subsequent finite-element analysis. In the present article the deep biomechanical analysis assuming creation of model in a finite-element complex is described.

In the dental practice even more often establish implants instead of traditional removable dentures. Application of implants has a number of advantages: there is no need to prepare basic teeth to establish a fixed denture; the patient easily adapts to a denture; use of such dentures doesn't cause problems unlike use of removable partial dentures; hygienic care of implants is simple, for this purpose there is a large number of

Z. Hu et al. (Eds.): AIPE 2021, LNDECT 119, pp. 181–195, 2022.
https://doi.org/10.1007/978-3-030-97064-2_18

accessories. Transitional implants establish at the initial stage of treatment for 4–6 months (the period of osteointegration of the permanent dental implants). Temporary fixed dentures allow to specify a final form of the finishing restoration, to create geometry of interdental nipples and harmonious occlusion relationship of teeth and tooth alignments [3]. In view of the fact that temporary bridge-like dentures are milled, after adaptation in an oral cavity they can be considered prototypes of permanent dentures.

In a number of works the research of a perspective of transitional implants installation [4–6] and clinical results of researches are described [7]. At implantation it is quite difficult for dentists to consider strength characteristics of a bone tissue, it is possible to estimate them enough approximately according to the tomogram. Modern methods of computer modeling on the basis of finite-element program complexes allow to create models of preliminary planning of operations for prosthetics and implantation by means of which it is possible to develop virtually a technique of transitional implantation and to receive recommendations about use of this or that model. It is possible to determine by a finite-element method zones of jaw parts with dental implants with high concentration of tension (these zones are inexpedient to involve at temporary and permanent prosthetics), i.e. to reveal schemes of designs, inapplicable in practice.

In the Russian and foreign sources it is published few works devoted to modeling of the stress-strain state of a design by a finite-element method at transitional implantation. In them problems of modeling of a separate implant in a jaw (without denture) are considered [8–10], or installation of implants on a toothless jaw with use of the simplified geometry of a denture is described [11, 12]. At the same time at a stage of treatment planning and control of implant osteointegration stomatologists even more often use a computer tomography which allows to create more difficult geometry of a design and to consider all stage of treatment planning, since installation of temporary dentures on transitional implants and finishing with fixing of permanent orthopedic designs.

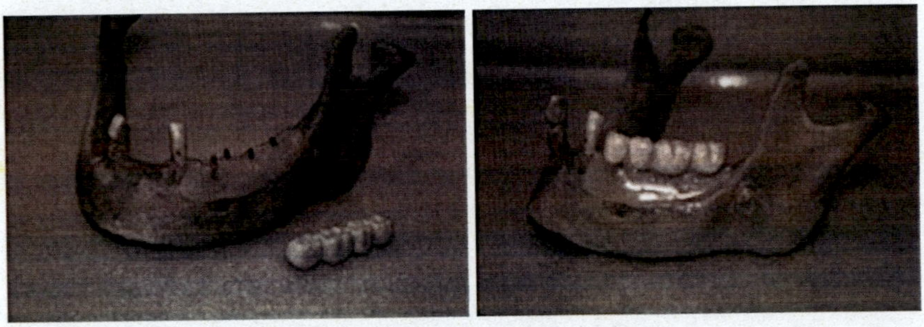

Fig. 1. Transitional implant-supported temporary bridge

Note that a temporary bridge is installed on a special type of flexible thermoplastic denture, made for a certain type of implant placement as Fig. 1 [13, 14].

2 Basic Relations of Continuum Mechanics

Within the framework of this work, we restrict ourselves to a brief presentation of the main ideas of the method necessary for understanding the research methodology and analyzing the results. A more detailed presentation of the issues under consideration is presented in special textbooks and monographs [15–18] and is beyond the scope of this study.

In mathematical modeling, teeth, cortical and cancellous bone, periodontium, prosthetic structure and other biological tissues are considered as a three-dimensional environment – a continuum endowed with certain mechanical properties. During the chewing process, the tooth tissues are exposed to external forces. In this case, the bone tissue is deformed, and internal stresses arise in it. Within the framework of this work, a traditional approach is used, which assumes that the hypotheses of continuity and homogeneity, generally accepted in the resistance of materials, can be used for bone tissue material.

The three-dimensional environment is represented as a set of material points, each of which occupies a position in space, specified by the initial coordinates (x, y, z) in the Cartesian coordinate system. When a solid is deformed, under the action of an external load, each material point is displaced to some neighboring point with coordinates (x', y', z'). In this case, the components of the displacement vector

$$\{u\} = \{u, v, w\}^T \tag{1}$$

are defined as the difference in the coordinates of the point before and after the application of the load

$$u = x' - x, v = y' - y, w = z' - z. \tag{2}$$

The displacements of the aggregate of material points lead to a change in the geometric shape of the body as a whole. Lateral $\varepsilon_x, \varepsilon_y, \varepsilon_z$ and shear $\gamma_{xy}, \gamma_{yz}, \gamma_{zy}$ deformations are a measure of the shape change in a small vicinity of the point under consideration. In the case of small deformations, deformations are associated with displacements by Cauchy dependencies.

$$\begin{aligned} &\varepsilon_x = \tfrac{\partial u}{\partial x}; \quad \varepsilon_y = \tfrac{\partial v}{\partial y}; \quad \varepsilon_z = \tfrac{\partial w}{\partial z}; \\ &\gamma_{xy} = \tfrac{\partial u}{\partial y} + \tfrac{\partial v}{\partial x}; \quad \gamma_{yz} = \tfrac{\partial v}{\partial z} + \tfrac{\partial w}{\partial y}; \quad \gamma_{zx} = \tfrac{\partial w}{\partial x} + \tfrac{\partial u}{\partial z}. \end{aligned} \tag{3}$$

In matrix form, the Cauchy relations are written as follows:

$$\{\varepsilon\} = [R]\{u\} \tag{4}$$

The deformation vector $\{\varepsilon\} = \{\varepsilon_x, \varepsilon_y, \varepsilon_z, \gamma_{xy}, \gamma_{yz}, \gamma_{zx}\}^T$ is expressed in terms of the displacement vector $\{u\}$ using a matrix of differential operators – $[R]$ – with the size (6×3) of the following form:

$$[R] = \begin{bmatrix} \frac{\partial}{\partial x} & 0 & 0 & 0 & \frac{\partial}{\partial z} & \frac{\partial}{\partial y} \\ 0 & \frac{\partial}{\partial y} & 0 & \frac{\partial}{\partial z} & 0 & \frac{\partial}{\partial x} \\ 0 & 0 & \frac{\partial}{\partial z} & \frac{\partial}{\partial y} & \frac{\partial}{\partial x} & 0 \end{bmatrix} \quad (5)$$

Note that three displacements uniquely determine six deformations, which indicates the need for the existence of three additional relations, known as the Codazzi-Gauss equations.

The stress state at a point of the body is written using 6 independent components of the stress state tensor, which are conveniently represented in the form

$$\{\sigma\} = \{\sigma_x, \sigma_y, \sigma_z, \tau_{xy}, \tau_{yz}, \tau_{zx}\}^T \quad (6)$$

In accordance with Hooke's law for an isotropic material, the relationship between the strain vector $\{\varepsilon\}$ and the stress vector is written in the form

$$\{\sigma\} = [D]\{\varepsilon\} \quad (7)$$

where $[D]$ is the matrix of elastic properties of the material.

$$[D] = \frac{E}{(1-\mu)(1-2\mu)} \begin{bmatrix} 1-\mu & \mu & \mu & 0 & 0 & 0 \\ \mu & 1-\mu & \mu & 0 & 0 & 0 \\ \mu & \mu & 1-\mu & 0 & 0 & 0 \\ 0 & 0 & 0 & \frac{1-2\mu}{2} & 0 & 0 \\ 0 & 0 & 0 & 0 & \frac{1-2\mu}{2} & 0 \\ 0 & 0 & 0 & 0 & 0 & \frac{1-2\mu}{2} \end{bmatrix} \quad (8)$$

The equilibrium conditions for an elementary parallelepiped cut out in the vicinity of the point under consideration can be written in the form.

$$\{R\}^T\{\sigma\} + \{b\} = 0, \quad (9)$$

here $\{\sigma\} = \{\sigma_x, \sigma_y, \sigma_z, \tau_{xy}, \tau_{yz}, \tau_{zx}\}^T$ is the stress vector, $\{b\} = \{b_x, b_y, b_z\}^T$ is the vector of external forces distributed over the volume of the body.

To obtain a closed system of equations, relations (9) are supplemented by the boundary conditions specified on the boundary of the body.

As a rule, on a part of the surface S_u, the boundary conditions are set in kinematic form

$$\{u\} = \{u\}^* \quad (10)$$

Here $\{u\}^*$ is the vector of specified displacements. On the rest of the body surface S_p, the boundary conditions are set in a static form:

$$\{p\} = [L]\{\sigma\} = \{p\}^*, \quad (11)$$

where $\{p\}^*$ is the vector of specified stresses, $[L]$ is the matrix of the direction cosines of the normal at the current point on the surface of the body.

The FEM is based on the idea of studying the behavior of a structure by analyzing the behavior of its individual parts $\Omega^{(e)}$, called finite elements.

The basic relations of the finite element method in the form of the displacement method can be obtained on the basis of the variational principle of the minimum of the total potential energy of the system, according to which the body is in equilibrium if the total potential energy of the system takes on a stationary value. In accordance with the main idea of the FEM, the desired displacement field in some subdomain $\Omega^{(e)}$ belonging to the body, called a finite element, can be found in the form:

$$\{u\} \approx [N]\{a\}^{(e)} \tag{12}$$

where [N] is a matrix of predefined basis functions, called form functions, $\{a\}^{(e)}$ is a vector of unknown displacements at certain points, called nodes of a finite element with volume $V^{(e)}$ and surface area $A^{(e)}$.

The use of the variational principle makes it possible to write down the equilibrium conditions of a finite element with a number – "*e*" in the form:

$$[K]^{(e)}\{a\}^{(e)} = \{F\}^{(e)} \tag{13}$$

where the matrix is $[K]^{(e)}$ called the stiffness matrix of the finite element, and $\{F\}^{(e)}$ is the vector of nodal forces for the element with a number "e", which are defined as:

$$[K]^{(e)} = \int_{\Omega^{(e)}} [B]^T[D][B]d\Omega^{(e)} \tag{14}$$

$$[F]^{(e)} = \int_{V^{(e)}} [N]^T\{b\}d\Omega^{(e)} + \int_{A^{(e)}} [N]^T\{p\}dA^{(e)} \tag{15}$$

here $[\mathbf{B}] = [R][N]$.

The stiffness matrix and the vector of nodal forces for an ensemble of finite elements are obtained by summing the corresponding values for individual finite elements using the formulas:

$$K_{ij} = \sum\nolimits_e K_{ij}^{(e)} \tag{16}$$

$$F_{ij} = \sum\nolimits_e F_{ij}^{(e)} \tag{17}$$

The final FEM resolving relations are represented in the form of a system of high-order linear algebraic equations:

$$[K]\{a\} = [F] \tag{18}$$

To solve system (18), special methods are used that take into account the specific features of the corresponding algebraic equations.

The above ideas and the general algorithm for constructing a mathematical model based on the finite element method were used in this work for the numerical analysis of a complex biomechanical system – a prosthetic structure, a tooth, implants and the bone tissue surrounding the implants.

When obtaining a three-dimensional image at the initial stage for a certain range of segmentation, we get tissue masks, which are then edited manually. Using the mask, you can build a surface mesh of the model.

The surface of the boundary of the regions is generally determined by an implicitly defined function. A triangulation obtained by one of the cellular methods is given for it. The most well-known cellular algorithms: the Caneiro method, the method proposed by Guezek, the Skala-method, the method of "Marching cubes", the method of "Marching tetrahedral", etc. We are setting an optimized mesh. By triangulation of a closed homogeneous boundary shell, a NURBS-surface is constructed for each body. Then an assembly unit is created with many inserts and many separate bodies.

It is often quite difficult to manually restore data from a tomographic image, in this case artificial intelligence systems for dental diagnostics are used, for example, the Diagnocat AI system [19]. Convolutional Neural Networks (CNN) are most commonly used for object detection and segmentation.

Segmentation is mainly used in separation of different tissues from each other, through extraction and identical features. One of the problems is the difficulty of reconstructing the model when identifying soft tissues, segmentation algorithms are constantly developing and improving [20].

Before recognizing the model, it is desirable to remove noise on the tomogram layers, visually the image becomes clearer [21]. When specifying algorithms for working with the tomogram image using the Matlab environment [22], you can increase brightness of images so that biologists can get accurate information from the images.

3 Temporary Denture Simulation

3.1 Purpose of Research

Main objective of the work – creation of biomechanical model of a jaw with the denture fixed on implants and the technique of planning of surgical treatment with use of transitional implants allowing carrying out simulation, to estimate results and to carry out the choice of optimum parameters of the future operation. It is necessary for achievement of this purpose: to carry out the analysis of the existing approaches to a solution of this problem, to develop a technique of forecasting of surgical treatment results at implantation operation, to estimate adequacy of the received results, to introduce results in practice of surgical treatment with use of transitional implants.

A method of using temporary prostheses based on temporary (d = 2.0 mm) dental implants, inserted into the lower jaw to a depth of 12–13 mm, and options for their placement are considered. Two-stage implants had dimensions for the molar d = 4.0 and 4.8 mm, and for the premolar d = 3.3.

For a number of reasons, there is often a problem of lack of space for correct positioning of the implant. The distance between the implants must be at least 2 mm. Prof. Arutyunov S.D. a variant of the arrangement of temporary implants along the diagonal was proposed [13, 14]. In this case, the actual problem of virtual modeling of the placement of implants, the configuration of the prosthesis and the assessment of the stress-strain state of the proposed structure.

Specialists identify the following problems of working with temporary implants [11]:

1) small thickness of bone tissue for installation;
2) it is difficult to install the implants in parallel (for example, an inclination of no more than 15^0);
3) it is difficult to assess the placement of implants on an orthopantomogram; for correct installation, you need to use computed tomography;
4) a small volume of bone in the posterior part of the upper jaw in the tuberous zone;
5) it is often necessary to install several small-diameter implants;
6) the implants are too short;
7) poorly adapted occlusion because the implants are loaded too early.

3.2 The Preliminary Biomechanical Analysis of Prototypes of the Fixed Bridge-Like Dentures Established on Transitional Implants

Let us introduce the following hypothesis: we believe that the loads on the denture are quite small (diet, soft type of food), the loads on the implant are small, and we investigate the stresses in the bone. In order to solve the problem of choosing a rational location for temporary implants, a simplified parametric model was introduced using a multi-agent approach (Figs. 2 and 3). The agent is a separate implant fixed in the jaw. The main characteristics of the agent were the geometric dimensions of the temporary implant, the depth and angle of its installation (Fig. 2). The angle varies within acceptable values. The structure of the bone tissue at the site of installation was characterized by the thickness of the cortical and cancellous bone layers, as well as by the physicomechanical characteristics of the bone tissue, determined from the diagnostic data. Based on the results of a preliminary calculation, the ultimate loads at the upper point of the implant were determined on the assumption that the latter has three translational and three rotational degrees of freedom. The implants are united by a suprastructure (Fig. 3), which implies joint loading of the entire prosthetic structure under the most unfavorable loading method. The search for a rational suprastructure was reduced to an optimization problem. The objective function was the achievement of the condition of equal strength in terms of stresses in the bone tissue in all possible places for the installation of transitional implants.

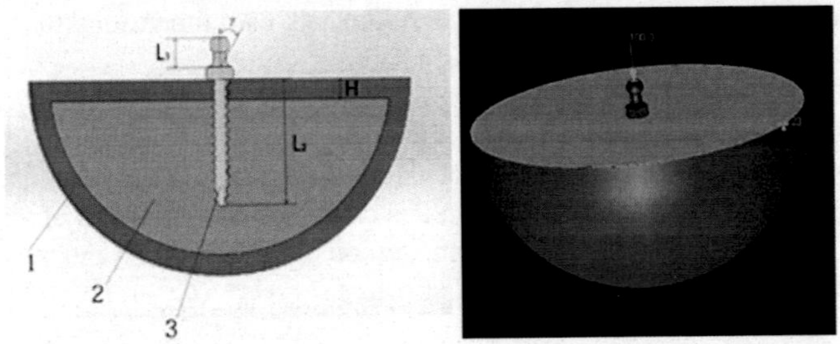

Fig. 2. "Transitional implant—jaw" model: 1—cortical bone (E^{cort}, ν^{cort}, σ_s^{cort}); 2—cancellous bone (E^{canc}, ν^{canc}, σ_s^{canc}); 3—implant (E^{imp}, ν^{imp}, σ_s^{imp})

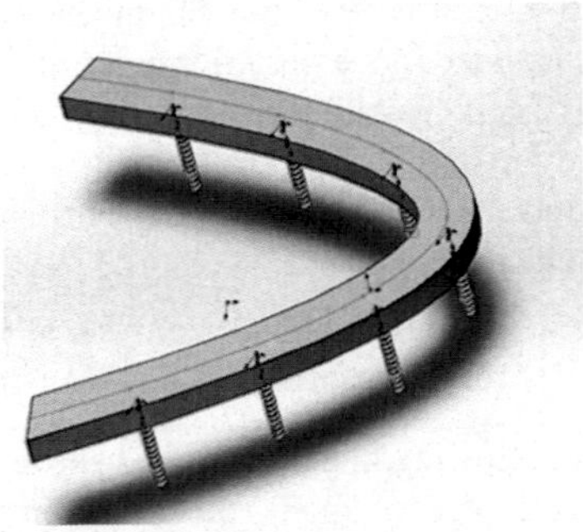

Fig. 3. Denture layout for implant placement with a multi-agent approach

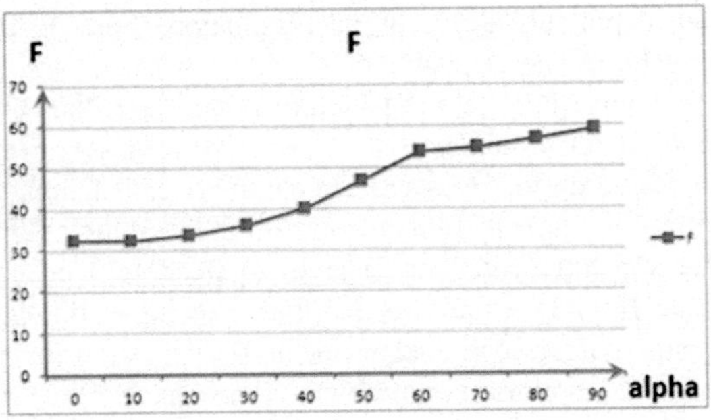

Fig. 4. Dependence of the ultimate load on the implant on the angle of inclination

Dependency plots for various load parameters and cortical bone thickness are shown in Figs. 4 and 5.

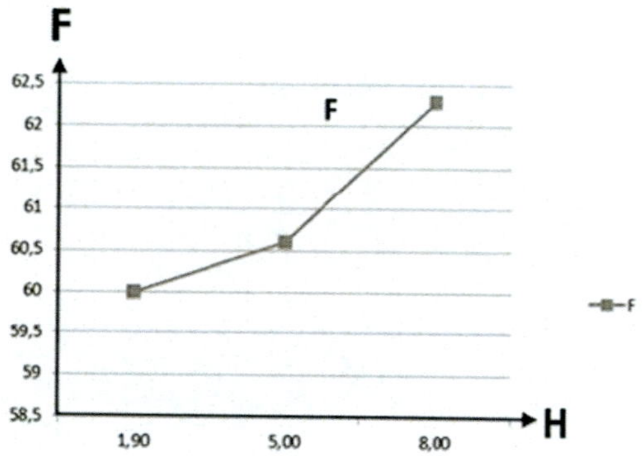

Fig. 5. Dependence of the ultimate load on the implant on the thickness of the cortical bone

In the model for a single implant were introduced the characteristics of bone density according to Misch [23] for four types of density. Density values can be obtained from a tomogram and determine approximately the type of bone. Table 1 shows approximate material characteristics for four types of bone according to [24].

Table 1. Material characteristics for different bone density

Type	E (MPa)	μ
D1	9500	0,3
D2	5500	0,3
D3	1600	0,3
D4	690	0,3

For the model of a temporary implant in the block a program was written in Python that allows positioning the implant and assessing the stress-strain state of the "implant-in-jaw" structure (Fig. 6).

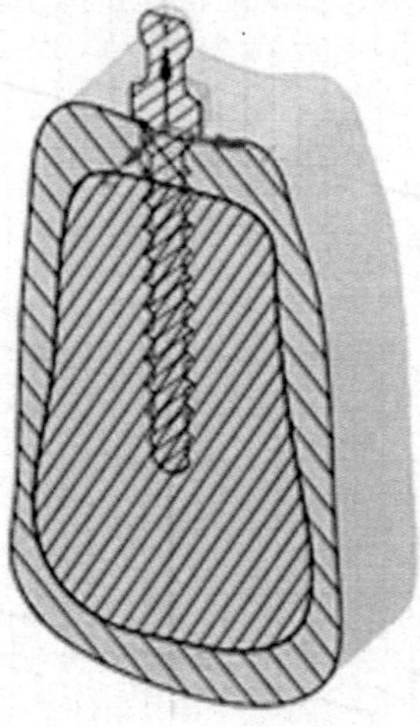

Fig. 6. Implant model in block

3.3 Biomechanical Analysis of a Full-Scale Model

On the example of the end defect of the dentition, a full-scale model was built using an individual computed tomogram. There was a preliminary assessment of the placement of two-stage and transitional implants and the preparation of data for the analysis of the strength characteristics of the planned orthopedic structures in order to study the distribution of functional loads. The options for calculating the models were considered using the example of the lower jaw with missing teeth 34, 35, 36, 37 and 38. Among the received placements for the calculation, only those spreads that can be applied in practice were selected.

Algorithm for creating a model:

1) Formation of a model of the jaw, teeth and denture by CT in the Mimics package (drawing a mask of objects in a certain density range, building a surface mesh of the model and its optimization). As a result, we get a surface mesh of the model, consisting of triangles.
2) Converting meshes in STL format to surfaces and solid models in SolidWorks using ScanTo3D for jaw, canine, denture.
3) Design of implants in the SolidWorks system.
4) Assembly of the resulting models of jaw, canine, denture and implants in SolidWorks.
5) Changing the geometry of the SolidWorks denture so that the it is supported on transitional implants (forming the “legs” of the denture).
6) Preparation of a computational model in Patran or Ansys (set material properties and boundary conditions of objects, generate a finite element mesh with a given quality and thickenings).

7) Calculation of stress, strain and displacement fields in the Nastran or Ansys system.
8) Structural failure assessment.

When obtaining a three-dimensional image in the Mimics software package, the following steps are performed:

1) Importing DICOM images.
2) Creating a segmentation mask: selecting a segmentation range (Thesholding).
3) Region Growing.
4) Mask editing.
5) Model reconstruction by mask.
6) Transforming the mesh in the Remesher module.
7) Export of the resulting model in STL format to SolidWorks.

First, using the described operations, we obtain a model of the jaw, then by editing the masks on the tomogram, masks of the denture, canine and jaw fragment are created, according to which surface meshes of 3D models are obtained.

When optimizing a surface mesh, the following steps are performed: Smoothing, Reduce, Auto Remesh, Quality Preserving Reduce Triangles.

The implants were designed in SolidWorks. The implant geometry is defined as a parameterized model with certain standard sizes.

The calculations were carried out in the MSC.Nastran system (Figs. 7 and 8). The properties of materials are given in Table 2. The values of von Mises stresses were obtained on average 0,8 MPa, which is 50% of the safety factor, i.e. the denture must be supported by the flexible thermoplastic covering prosthesis.

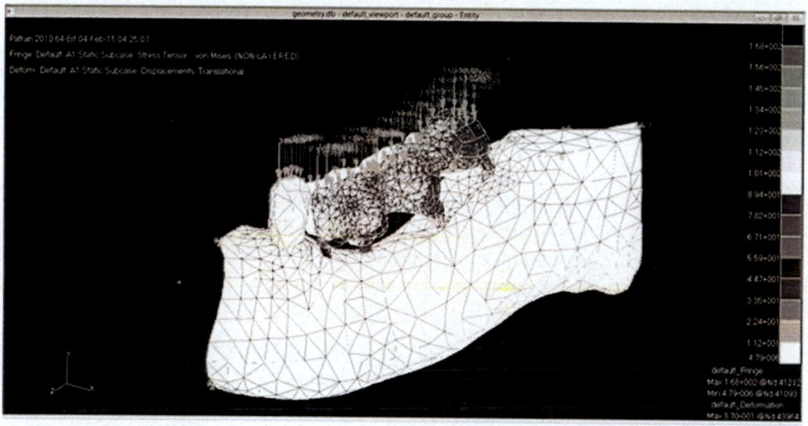

Fig. 7. Von Mises stress field for a model of four transitional implants

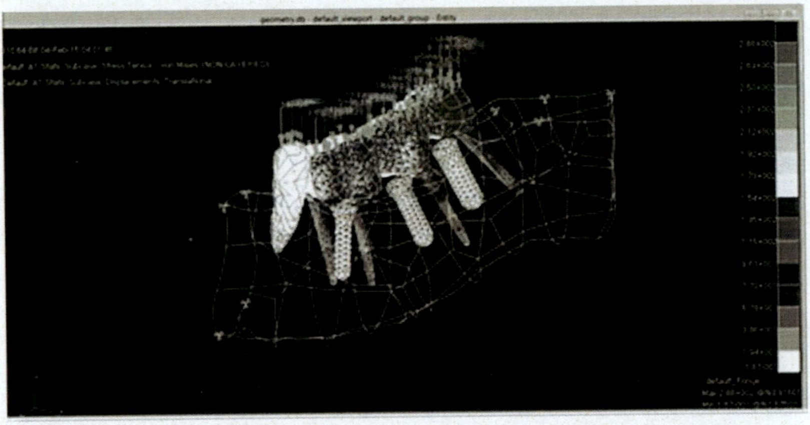

Fig. 8. Displacement field of the structure for a model of four transitional implants, three two-stage implants in the case of a maximum length of 18 mm small-diameter implants

Table 2. Material characteristics for model

Material	Young's modulus, MPa	Poisson's ratio, б/р	Strength, MPa	
			Compression	Tension
Dentine	14700	0,31	167	55
Cancellous bone	7500	0,45	82	15
Polymer	2600	0,33	33	30
Titanium grade 4	117000	0,33	280	600

On the obtained geometric model, several variants of placement of implants in the jaw were previously considered, taking into account the condition that the minimum distance between the implants is 2.0 mm. At the same time, it was believed that if a transitional implant cannot be positioned in a row with a two-stage one, then a diagonal arrangement between them is possible. As a result, it was found that no more than three two-stage and four transitional implants can be installed on this fragment of the jaw.

To study the dependence of stresses on the length of implants, lengths of 13 and 16 mm were chosen as the most common in practice. 1st option: increase the length of the cylinder to 12.5 mm, do not change the length of the cone (3.5 mm). 2nd option: increase the length of the cylinder to 10.5 mm, the length of the cone to 5.5 mm. In the case of elongation of the cone, the maximum von Mises stresses are significantly reduced. However, compared with previous calculations, stresses increase for an implant with an intraosseous part length of 13 mm.

A physical prototype was made for one of the models using a 3DSYSTEMS ZPrinter ® 650 3D printer, see Fig. 9.

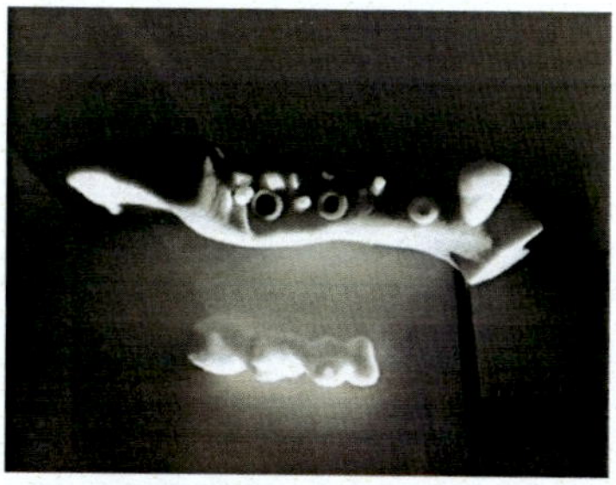

Fig. 9. A model printed on a three-dimensional printer

As part of the multi-agent approach, a variant of optimizing the calculations of the position of transitional dental implants in the jaw was proposed. On a simpler parametric model with the introduction of basic parameters, it is possible to carry out a basic series of calculations for the proposed task of modeling the placement of implants and choose the most suitable solutions within the criterion that all implants should evenly distribute stresses in the bone. At the final stage of modeling, a full-scale model of the patient's tomogram is designed, on which a specific calculation result for a specific clinical case can be verified. Diagnostics using this approach will allow taking into account the strength characteristics of bone tissues and predicting the durability of the functioning of temporary dentures.

Analysis of the technique of temporary prosthetics has shown that in most cases, special-type temporary flexible thermoplastic denture have more advantages compared to the use of transitional implants. It is also worth noting that currently a one-stage implantation technique with immediate loading of the structure has become more actual.

4 Summary and Conclusion

The received models allow to estimate loads of an implant and an denture as on schematical model (on which it is possible to receive express estimates), and on the individual, constructed according to the tomogram, full-scale model.

On the basis of the researches conducted in this work it is possible to draw the following conclusions:

1) Experimental data on properties of jaw bone tissues are analysed and generalized.
2) The model allowing giving express estimates of a condition of bone tissues after loading is developed to carry out the analysis of distribution of tension to bones and to optimize arrangement of implants.
3) The technique of creation of denture's biomechanical model with a support on transitional implants taking into account specific features of density of bone tissues and the geometry received according to the tomogram is developed.

4) The research of the stress-strain state of the jaw's part making on the example of terminal defect of a tooth alignment with a denture on transitional implants on the basis of a finite-element method is conducted.
5) The design of a transitional implant taking into account selection of the carving depth and the head form is improved.
6) Two approaches to modeling of arrangement of transitional implants are entered: on simpler model with in advance set geometry optimization is considered; on full-scale model total calculation was carried out.
7) During the work on denture formation it was used the developed program complex for biomechanical facet models modified under process of denture formation with a support on transitional implants.
8) The physical prototype of model by means 3D-printer has been made.

Acknowledgment. Work is performed with assistance of the RFBR № 16-07-01100 "The development of the theoretical bases of intelligent simulation of temporary implants' positioning at a two-stage technique of implantation".

References

1. Matveeva, A.I., Ivanov, A.G., Gvetadze, R.S., Gavryushin, S.S., Karasyov, A.V.: Improving the effectiveness of orthopedic treatment of patients on the basis of mathematical simulation of promising implant designs. Stomatology **76**(5), 44–48 (1997). (in Russian)
2. Matveeva, A.I., Kanatov, V.A., Gavryushin, S.S.: The use of mathematical modeling in the improvement of orthopedic treatment of end defects in the dentition. Stomatology **69**(1), 48–51 (1990). (in Russian)
3. Arutyunov, S.D., Eroshin, V.A., Perevezentseva, A.A., Boyko, A.V., Shirokov, I.Y.: Criteria for the strength and durability of temporary fixed dentures, vol. 4, no. 89, pp. 84–85. The Dental Institute (2010). (in Russian)
4. Froum, S., Emtiaz, S., Bloom, M., Scolnick, J., Tarnow, D.: The use of transitional implants for immediate fixed temporary prostheses in cases of implant restorations. Pract. Periodontics Aesthet. Dent. **10**(6), 737–746 (1998)
5. Dilek, O., Tezulas, E., Dincel, M.: Required minimum primary stability and torque values for immediate loading of mini dental implants: an experimental study in nonviable bovine femoral bone. Oral Surg. Oral Med. Oral Pathol. Oral Radiol. Endodontol. **105**(2), 20–27 (2008)
6. Scepanovic, M., et al.: Immediately loaded mini dental implants as overdenture retainers 1-year cohort study of implant stability and peri-implant marginal bone level. Ann. Anat. **199**, 85–91 (2015)
7. Robustova, T.G., Put, G.A.: The use of transitional intraosseous dental implants. Russ. J. Dent. **1**, 46–49 (2005). (in Russian)
8. Jayaraman, S., Mallan, S., Rajan, B., Anachaperumal, M.P.: Three-dimensional finite element analysis of immediate loading mini over denture implants with and without acrylonitrile O-ring. Indian J. Dent. Res. **23**(6), 840–841 (2012)
9. Arutyunov, S.D., Panin, A.M., Antonik, M.M., Iun, T.E., Adamian, R.A., Shirokov, I.Y.: Stomatologiya — Stomatology, 1(91), 54–58 (2012). (in Russian)

10. Hasan, I., Heinemann, F., Aitlahrach, M., Bourauel, C.: Biomechanical finite element analysis of small diameter and short dental implant: extensive study of commercial implants. Biomed. Tech. (Berl) **57**(1), 21–32 (2012)
11. Fatalla, A.A., Song, K., Du, T., Cao, Y.: a Three-dimensional finite element analysis for overdenture attachments supported by teeth and/or mini dental implants. J. Prosthodont. **21** (8), 604–613 (2012)
12. Bullis, G., Golestanian, V.: Predicting the performance of mini implant-retained prostheses using finite element analysis. Incl. Restor. DrivenImplant Solutions **2**(1), 24–29 (2011)
13. Arutyunov, S.D., et al.: A method of temporary prosthetics with fixed dental bridges on dental implants. Patent 2432924. Russian Federation, bulletin no. 31, vol. 3, p. 698. (in Russian)
14. Arutyunov, S.D., et al.: Temporary fixed denture. Patent 2494702, Russian Federation (2013). (in Russian)
15. Chumachenko, E.N., Arutyunov, S.D., Lebedenko, I.Y.: Mathematical Simulation of the Stress-Strain State of Dentures, p. 272. Molodaya gvardiya (2003). (in Russian)
16. Dimitrienko, Yu.I.: Continuum mechanics: in 4 volumes. Fundamentals of solid state mechanics, vol. 4, p. 623. Bauman Moscow State Technical University (2013). (in Russian)
17. Galanin, M.P., Savenkov, E.B.: Methods of Numerical Analysis of Mathematical Models, 2nd edn, p. 591. Bauman Moscow State Technical University (2018). (in Russian)
18. Koltunov, M.A., Kravchuk, A.S., Mayboroda, V.P.: Applied mechanics of deformable solids. – M.: Higher School, p. 349 (1983). (in Russian)
19. Ezhov, M., et al.: Clinically applicable artificial intelligence system for dental diagnosis with CBCT. Sci. Rep. **11**, 15006 (2021). https://doi.org/10.1038/s41598-021-94093-9
20. Kama, R., Chinegaram, K., Tummala, R.B., Ganta, R.R.: Segmentation of soft tissues and tumors from biomedical images using optimized k-means clustering via level set formulation. Int. J. Intell. Syst. Appl. (IJISA) **11**(9), 18–28 (2019). https://doi.org/10.5815/ijisa.2019.09.03
21. Hamdi, M.A.: A comparative study in wavelets, curvelets and contourlets as denoising biomedical images. Int. J. Image Graph. Sig. Process. (IJIGSP) **4**(1), 44–50 (2012). https://doi.org/10.5815/ijigsp.2012.01.06
22. Sharma, G.: Performance analysis of image processing algorithms using matlab for biomedical applications. Int. J. Eng. Manuf. (IJEM) **7**(3), 8–19 (2017). https://doi.org/10.5815/ijem.2017.03.02.10.5815/ijem.2017.03.02
23. Misch, C.E.: Dental Implant Prosthetics, p. 637. Elsevier/Mosby, Amsterdam (2005)
24. Premnath, K., Sridevi, J., Kalavathy, N., Nagaranjani, P., Sharmila, M.R.: Evaluation of stress distribution in bone of different densities using different implant designs: a three-dimensional finite element analysis. J. Indian Prosthodont. Soc. **13**(4), 555–559 (2012). https://doi.org/10.1007/s13191-012-0189-7

Methodology of a Comprehensive Assessment of the Potential of Regional Energy

Nataliya Mutovkina(✉)

Tver State Technical University, Tver 170012, Russia

Abstract. The socio-economic development of the regions is unthinkable without continuous provision of the energy generation system with energy resources. At the same time, it is necessary to be guided by energy conservation, resource renewal, and environmental safety. Any changes to the existing energy system in the region should be justified and carried out after a systematic analysis and assessment of the potential of regional energy. A system analysis and evaluation of the possibility of regional energy allow us to identify the shortcomings of the existing energy system and identify opportunities for the construction of new thermal power plants and technical re-equipment and modernization of regional energy facilities. The article discusses a methodology for assessing the potential of regional energy based on a systematic approach. The process involves the assessment of both traditional energy sources in the region and renewable "green" energy sources. System analysis, as the primary research method, allows us to identify regional features of the socio-economic sphere, the location of settlements and infrastructure facilities, the formed energy system, natural conditions. The proposed methodology is universal and can be used to assess the potential of the energy system of any region. The practical value of the study lies in the application of the proposed methodology will allow the leadership of the area to make more rational and effective decisions regarding the development of regional energy.

Keywords: The region's energy system · Energy potential of the area · Alternative energy sources · "Green" energy · System analysis · Expert assessments

1 Introduction

The formation of the optimal composition, structure, energy capacity, and other parameters of the energy system of the region, as well as the permanent provision of its working condition, is the essential condition for creating decent living conditions for the population of the area and the development of the regional economy.

However, optimizing the energy system is a complex, ambiguous process requiring significant material, financial, labor, and time costs. In addition, it should be borne in mind that one of the areas of optimization of the regional energy system in the construction and modernization of thermal power plants using so-called "green" energy sources. Therefore, the analysis of the possibilities of using renewable "green" energy sources in the region is one of the most critical sections of the proposed methodology.

Z. Hu et al. (Eds.): AIPE 2021, LNDECT 119, pp. 196–206, 2022.
https://doi.org/10.1007/978-3-030-97064-2_19

Thus, the study aims to develop a methodology for a comprehensive assessment of the region's energy potential, designed to justify the costs of upgrading the existing energy system. Identifying the possibilities of using renewable "green" energy sources in the region is an essential element of the novelty of the proposed methodology, which has practical value. Moreover, the use of renewable "green" energy sources allows obtaining significant economic and environmental effects.

1.1 Background and Relevance of the Study

A system analysis of the potential of regional energy makes it possible to assess the feasibility of building thermal power plants operating on alternative energy sources. This is especially true for unevenly populated regions, where recreational resources, rare species of plants and animals are located.

A comprehensive assessment of regional energy potential includes:

1. An analysis of natural and economic conditions.
2. The location of thermal and electric power consumers in the region.
3. An evaluation of the energy capacities available in the area.
4. An assessment of the state of energy facilities.

Analyzing the possibilities of using traditional and alternative energy sources makes it possible to determine the optimal combination of thermal power plants operating on both energy sources.

Energy generators powered by alternative energy sources are indispensable in small settlements, farms, and small enterprises that are significantly remote from sizeable thermal power plants.

1.2 Literature Review

Since the analysis of the possibilities of using alternative energy sources in the regional energy system is one of the main sections of the methodology, publications on renewable sources were considered.

Such sources include wind and water energy, solar energy, geothermal energy, biofuels. However, the use of "green" energy sources also contains specific problems. Firstly, there is not enough solar, geothermal, and wind activity everywhere, and the location of full-flowing rivers is entirely not uniform. Furthermore, the burning of biofuels entails significant emissions of carbon dioxide and other harmful substances into the atmosphere, so the use of biofuels is possible only if there are technologies that provide for multi-level filtration. But such technologies are costly. Secondly, the economic feasibility of developing and placing thermal power plants on alternative energy sources is questioned. For example, the world's largest wind farms have low profitability, and some are recognized as unprofitable. This is because, during the service life (approximately 20 years), the wind turbine does not have time to generate enough energy to justify the costs of its assembly, installation, and maintenance [1].

However, wind generators fully justify their use with sufficient wind speed and the absence of temperature differences [2–4]. The same can be said about other alternative energy sources.

The research was based on the work of Russian and foreign scientists in the field of assessing the possibilities of using renewable energy sources [3, 5–12], the chances of modernization and optimization of the energy complex of the region [13–17], as well as publications on the theory of systems and system analysis, concepts of "green" energy, resource conservation [18–25]. For example, in [3], the problem of monitoring the consumption of electricity generated by solar panels is considered. It is proposed to introduce an automated information system for monitoring the operation of this system to reduce excessive energy losses. The paper [11] talks about the possibilities of overcoming the energy crisis with solar energy. Of course, it is advisable to build power plants based on solar energy for regions with such solar activity as in Pakistan. The use of wind power plants is discussed in detail in [12], and the importance of using alternative energy sources is also discussed.

The issues of energy conservation and saving of energy resources are considered in such works as [23–25].

The analysis of the reviewed publications allowed us to come to an unambiguous conclusion: power plants using renewable energy sources are quite expensive and complex, but they have several advantages over traditional energy. The main benefits are independence from fossil fuels and low environmental damage. In addition, the development of renewable energy stimulates innovative production and makes it possible to provide isolated areas and settlements with electricity [8, 10, 27]. These arguments determine the need for a systematic analysis of the potential of regional energy to identify the possibility of using renewable energy sources.

2 Methodology of Comprehensive Assessment of Regional Energy Potential

2.1 Assessment of Available Energy Generation Capacities in the Region

To assess the energy generation capacities currently available in the region, it is necessary to consider the dynamics of electricity generation over equal periods (for example, by year) in connection with the volume of electricity consumption. Thus, it is possible to assess the region's needs for additional generating capacities [6, 13, 17].

Next, the energy security of the region is assessed, namely, the presence of traditional energy sources in the area, the dependence of the region on imported fuels.

The structure of regional energy is considered: the number of power plants and their total capacity, a grouping of power plants by type [17]. Here it seems appropriate to build pie charts. The estimation of power generation capacities can also be presented in the form of a table. It should indicate the list of power plants and heat generation stations, their current state, capacity (megawatts), and the share each occupies in the total generation (in %).

A separate stage of the assessment is analyzing the technical condition of thermal power facilities: the percentage of depreciation, renewal, and disposal of fixed assets.

2.2 Analysis of the Level of Provision of the Region with Energy Resources

At this stage, the distribution of the population across the region and the location of existing energy facilities are considered. The issue of the degree of availability of heat and energy for the region's people is being resolved. Statistical information is collected on the proportion of the population remaining without electricity and heat supply. The coefficient of provision of the region with heat and electric energy is calculated as the ratio of the number of provided districts to the total number of sections of the area. The higher the value of this indicator, the better the region is supplied with energy resources.

A cartogram is being drawn upon which the objects of the regional energy system planned for commissioning, including energy facilities operating on replenished "green" sources, are plotted. The planned security coefficient is calculated.

Both values are compared. If the planned value significantly exceeds the actual value, a decision is made to introduce additional capacity.

2.3 Analysis of "Green" Resources for the Production of Heat and Electricity in the Region

This type of analysis is performed for each alternative source, briefly discussed in Sect. 1.2 of this article.

The conversion of airflow energy into electricity is possible only in a specific range of wind speeds. Therefore, it is also essential to determine what percentage of the wind speed decreases below the critical values [26]. The boundary of maximum wind speeds for the territory has a unique, limiting value. For various designs of wind turbines, the upper limit of the operating range lies at the level of 20 to 45 m per second. If these values are exceeded, the design of the windmill is in danger of destruction [27].

The primary indicator is the level of solar activity [7, 15]. In addition, the feasibility of building geothermal power plants is determined by indicators such as the temperature of vapor, thermal power, and subthermal waters at the river's mouth.

The initial information for assessing the fuel potential of solid biofuels from plant and wood waste is information on the gross harvest of crops and the volume of harvested wood, the proportion of the waste generated. For example, the assessment can be carried out by the methodology proposed in [28]:

1) The amount of waste from crops, logging, and woodworking is calculated according to the formulas:

$$W_1 = V \cdot \alpha_1, W_2 = V_2 \cdot \alpha_2, W_3 = V_2 \cdot \alpha_3 \quad (1)$$

V_1 is the gross harvest of crops (thousand tons), V_2 is the volume of harvested wood (thousand tons), α_1, α_2, and α_3 share crop waste, logging, and woodworking, respectively.

Then the total amount of waste will be:

$$W = W_1 + W_2 + W_3 \tag{2}$$

2) The energy content of a fuel is calculated in tons of conventional fuel per ton (t.c.f./t) according to the formula:

$$E = \frac{Q_1}{Q_2} \tag{3}$$

Q_1 is the calorific value of natural fuel (MJ/kg or kcal/kg), Q_2 is the calorific value of a ton of conventional fuel (MJ/kg or kcal/kg). A ton of traditional fuel is a unit of energy measurement equal to $2.93 \cdot 10^{10}$ Joules. It is defined as the amount of energy released during the combustion of one ton of fuel with a calorific value of 7 000 kcal/kg, which corresponds to the typical calorific value of coal.

3) The fuel potential of agro pellets from agricultural waste (P_1), pellets from wood chips, wood pellets, bio-coal from wood waste (P_2) in thousand tons of conventional fuel is calculated by the formulas:

$$P_1 = W_1 \cdot E_1, \quad P_2 = (W_2 + W_3) \cdot E_2 \tag{4}$$

The calorific value and energy content of natural fuel (Q_1) are known for different fuel types. Their normative values can be found in the specialized literature, for example, in [28].

4) Technical electric power (P_3, in a million kWh) and thermal energy (P_4, in a million Gcal) potentials of agro pellets are determined by the formulas:

$$P_3 = 8141 \cdot P_1 \cdot \beta_1 \cdot (1 - \gamma_1)/1000, \; P_4 = 7 \cdot P_1 \cdot \beta_2 \cdot (1 - \gamma_2)/1000 \tag{5}$$

The constant 8 141 means the conversion of one ton of conventional fuel into kWh; β_1 is the conversion coefficient into electricity of mini-thermal power plants with direct combustion of biomass; γ_1 is the share of electricity for own needs; β_2 is the conversion coefficient into thermal energy of mini-thermal power plants with direct combustion of biomass; γ_2 is the share of thermal energy for own needs. The values of β_1, β_2, γ_1, and γ_2 are set expertly.

5) Technical electric power (P_5), in a million kWh) and thermal power (P_6, in a million Gcal) the potentials of wood chips, wood pellets, bio-coal are calculated according to the formulas:

$$P_5 = 8141 \cdot P_2 \cdot \beta_3 \cdot (1 - \gamma_1)/1000, \; P_6 = 7 \cdot P_2 \cdot \beta_4 \cdot (1 - \gamma_2)/1000 \tag{6}$$

β_3 is the conversion coefficient to the electricity of mini-thermal power plants; β_4 is the conversion coefficient to the thermal energy of a boiler house or mini-thermal power plant. The values of β_3, β_4 are also set by experts.

A particular type of biofuel is biogas, which usually consists of methane (55–85%) and carbon dioxide (15–45%). In addition, it may contain hydrogen sulfide. The heat of combustion of biogas ranges from 21.0 to 27.2 MJ/m^3. According to the heat of combustion, 1 m^3 of biogas is equivalent to 0.8 m^3 of natural gas, 0.7 kg of fuel oil, 0.6 kg of gasoline, 1.5 kg of absolutely dry firewood, 3 kg of manure briquettes. When processing 1 ton of solid household waste of organic origin, it is possible to obtain from 45 to 100 m^3 of biogas. Thus, 1 ton of fresh waste from cattle and pigs at 85% humidity gives from 45 to 60 m^3 of biogas, 1 ton of chicken manure at 75% humidity allows you to get up to 100 m^3 of biogas [29].

Therefore, if animal husbandry is sufficiently developed in the region, biogas production can be considered one of the solutions to the problem of using traditional energy sources.

2.4 Decision-Making Based on the Results of the Analysis

At the final stage of the analysis of regional energy potential, the results of the previous steps are summarized. Indicator tables are being filled out showing the presence of traditional energy sources in the region, the state of fixed production assets in the energy sector, the availability of various types of energy to the population, and the possibility of using alternative "green" energy sources.

All the listed tables are presented in the third section of this article. The rating of each key indicator has five gradations: "low," "below average," "average," "above average," and "high." Experts set the interval values for each gradation, taking into account the peculiarities of the considered branch of the economy and regional peculiarities.

3 The Potential of the Energy System of the Tver Region

By the presented methodology, a comprehensive assessment of the potential of the energy system of the Tver region was carried out.

The dynamics of electricity generation and consumption in the Tver region are shown in Fig. 1. The Tver region supplies electricity to neighboring areas since it produces electricity many times more than it consumes.

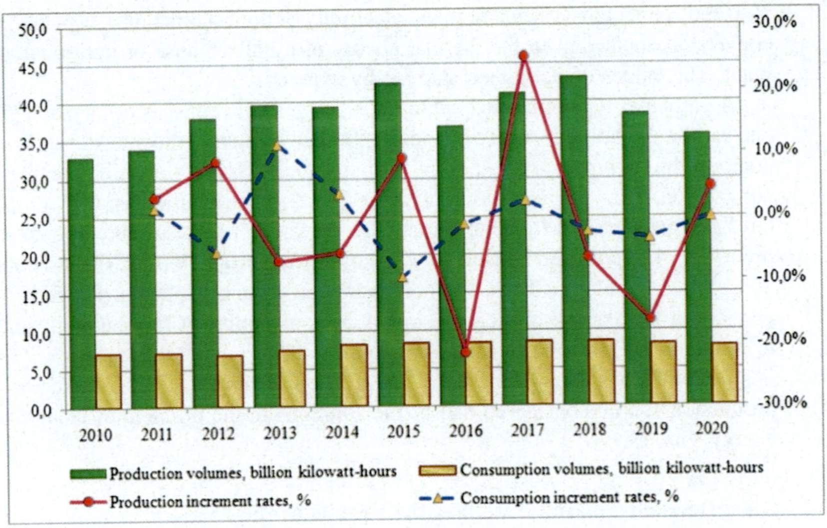

Fig. 1. Indicators of the dynamics of electricity generation and consumption

At the same time, the region is highly dependent on imports of traditional energy sources (Fig. 2).

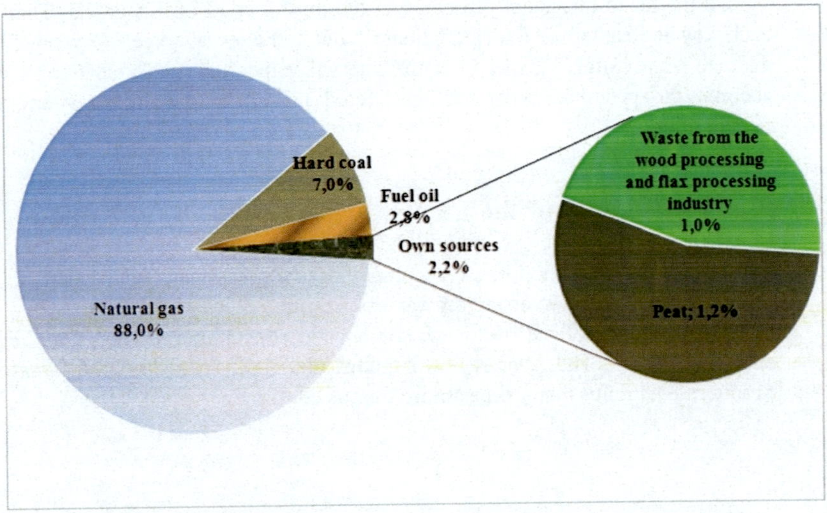

Fig. 2. Structure of imported and own energy sources in the Tver region

The Kalinin nuclear power plant occupies the leading share in the energy structure of the Tver region (Fig. 3), which threatens the life and health of the population and the preservation of local flora and fauna.

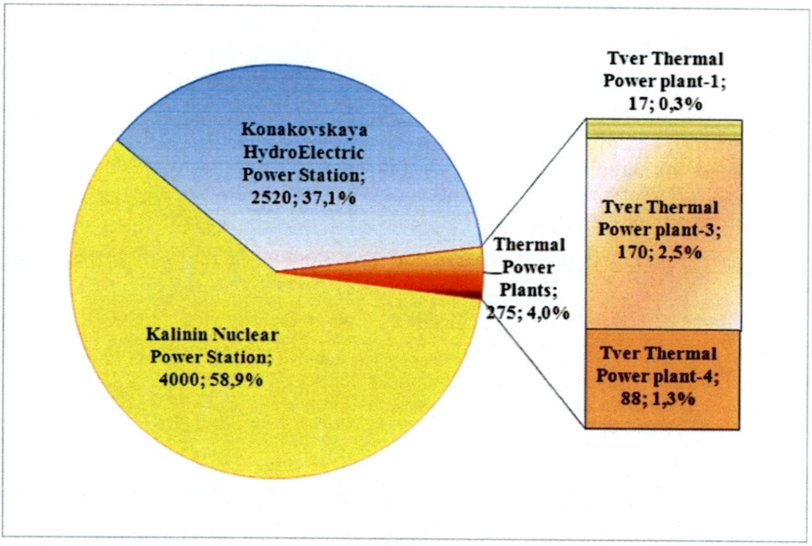

Fig. 3. Energy structure of the Tver region

The results of the analysis of the first stage of the methodology are presented in Fig. 4.

Dynamics of energy production and consumption in the region

	Dynamics of Production (**DP**)	Dynamics of Consumption (**DC**)
	high	high
Ratio	**DP>>DC**	

Availability of traditional energy sources in the region

Natural gas	Hard coal	Oil and petroleum products	Water	Peat
–	–	–	**above average**	**high**

The state of fixed production assets in the energy sector of the region

Percentage of wear	Percentage of updates	Percentage of disposal
45,7	4,2	0,4
high	**low**	**low**

Fig. 4. Results of the first stage of the analysis

Tver region imports all traditional energy sources while possessing rich peat reserves. Since the timber processing industry is developed in the region, logging and woodworking waste are also available. The development of the agricultural sector is the reason for the increase in the share of waste from crop production and animal husbandry.

The population of the Tver region is 1 245 619 people as of 01.01.2021. The population density is 14.79 people per square kilometer. The urban population is 77.04%. Its energy supply is 100%. However, the population is highly unevenly distributed throughout the region. Despite the excess capacity of thermal and electric power generation facilities, many rural settlements remain without these benefits of civilization. This is one of the reasons people move to cities. The average coefficients of energy supply to the population and infrastructure facilities are presented in Table 1.

Table 1. Energy security

Security coefficients	Actual values	Planned values	Variation	Conclusion
Electricity	0.98	1.00	−0.02	High
Thermal energy	0.74	1.00	−0.26	Above average
Gas	0.64	1.00	−0.36	Average
Average coefficient	0.78	1.00	−0.22	Above average

Table 2 shows the calculations and assessment of the fuel potential of the Tver region of solid biofuels from plant and wood waste.

Table 2. Assessment of the energy potential of biofuels

Type of potential	Calculated values	Conclusion
P_1, thousands of tons of conventional fuel	39.03	Average
P_2, thousands of tons of conventional fuel	1351.81	Above average
P_3, millions of kilowatt-hours	75.46	Above average
P_4, millions of gigacalories	0.19	Low
P_5, millions of kilowatt-hours	3659.20	Above average
P_6, millions of gigacalories	7.19	Average

An analysis of the possibilities of using alternative "green" energy sources in the Tver region showed that biofuels could be used from all sources. Its volumes in the region are above average. In addition, the number of livestock farm animals in the Tver region makes it possible to obtain biogas for electricity generation.

4 Summary and Conclusion

Thus, the assessment of the energy capacities available in the region makes it possible to identify the need for the introduction of additional capabilities or to state their sufficiency. The analysis of the area's energy security level includes an assessment of the technical condition of power facilities and the region's dependence on imported fuels. Comparing the costs of importing and developing energy sources available in the area makes it possible to decide on the further development of the regional energy system.

The methodology provides for the calculation of relative characteristics. The initial information for analyses is the data of official statistics. Then, the results of the calculations are compared with the expert-defined intervals, the boundaries of which may vary depending on regional characteristics and depending on the degree of consistency of expert opinions.

The application of the developed methodology is shown by assessing the energy potential of one of the regions of the Russian Federation – the Tver region.

According to the assessment results, it was determined that in the Tver region, the largest share of installed capacities falls on nuclear generation, thermal generation is in second place in importance. Assessment of the potential of renewable energy sources in the Tver region has shown that the placement of generation facilities using solar, wind, geothermal and small river energy is impractical. The assessment of the bioenergy potential of the livestock, logging, and timber processing industries showed that the development of biogas energy is relevant in the Tver region. This, on the one hand, will allow waste disposal, and on the other hand, will provide electricity and heat to small settlements.

References

1. The problem of wind turbine payback: dimensions, installation, operation of the windmill (2021). https://energo.house/veter/okupaemost-vetrogeneratora.html. Accessed10 Oct 2021
2. Haapala, K.R., Prempreeda, P.: Comparative life cycle assessment of 2.0 MW wind turbines. Int. J. Sustain. Manuf. **3**(2), 170–185 (2014)
3. Praiselin, W.J., Belwin, E.J.: A review on impacts of power quality, control and optimization strategies of integration of renewable energy based microgrid operation. Int. J. Intell. Syst. Appl. (IJISA) **10**(3), 67–81 (2018)
4. Tiwari, R., Babu, N.R.: Comparative analysis of pitch angle controller strategies for PMSG based wind energy conversion system. Int. J. Intell. Syst. Appl. (IJISA) **9**(5), 62–73 (2017)
5. Esyakov, S.Y., Lachuga, Y.F., Varfolomeev, S.D., Red'ko, I.Y., Vasil'ev, A.N.: The potential use of renewable energy in agriculture. Plumb. Heat. Air Cond. **1**(169), 88–92 (2016). (in Russian)
6. Samarina, V.P., Skufina, T.P.: Experience and prospects of using alternative energy sources in the energy-surplus region of Russia. North Mark. Form. Econ. Order **5**(56), 190–198 (2017)
7. Singh, A.K., Kumar, M., Kumar, D., Singh, S.N.: Heterostructure silicon and germanium alloy based thin film solar cell efficiency analysis. Int. J. Eng. Manuf. **10**(2), 29–40 (2020)
8. Grosheva, E.K., Chuprina, A.D.: Alternative energy sources and their application in Russia. Bus. Educ. Knowl. Econ. **1**(15), 19–23 (2020)
9. Zaika, V.A., Pchelkin, V.O., Filimonov, D.V.: Problems and prospects of financing the development potential of alternative energy in Russia. Actual Prob. Mil. Sci. Res. **S3**(15), 124–134 (2021)
10. Alternative energy: prospects for the development of the renewable energy market in Russia (2021). https://delprof.ru/press-center/open-analytics/alternativnaya-energetika-perspektivy-razvitiya-rynka-vie-v-rossii/. Accessed 11 Oct 2021
11. Haq, Q.A.U.: Design and implementation of solar tracker to defeat energy crisis in Pakistan. Int. J. Eng. Manuf. (IJEM) **9**(2), 31–42 (2019)

12. Hemanth Kumar, M.B., Saravanan, B.: Power quality improvement for wind energy conversion system using composite observer controller with fuzzy logic. Int. J. Intell. Syst. Appl. (IJISA) **10**(10), 72–80 (2018)
13. Oleynikova, E.V., Kolotienko, M.A., Danilenko, I.Y., Kovalev, V.V.: Assessment of the resource potential of the Southern Federal District for the construction of innovative solar and wind power plants. Eng. Bull. Don **2**(49), 11–17 (2018)
14. Kurmangalieva, A.R.: Engineering and environmental restrictions in the construction of solar power plants. Geol. Geogr. Global Energy **4**(71), 12–17 (2018)
15. Nguyen, T.N., Sizov, V.D., Vu, M.P., Cu, T.T.H.: Evaluation of work efficiency of the solar power plant installed on the roof of a house in Hanoi city. Energetika Proc. CIS Higher Educ. Inst. Power Eng. Assoc. **63**(1), 30–41 (2020). https://doi.org/10.21122/1029-7448-2020-63-1-30-41. (in Russian)
16. Latif, S.D.: Technical improvement of air pollution through fossil power plant waste management. Int. J. Eng. Manuf. **10**(4), 43–53 (2020). https://doi.org/10.5815/ijem.2020.04.04
17. Dolinskaya, I.M., Yakovenko, E.M.: Prospects for the development of renewable energy sources in the territorial planning of the Magadan region. Universum: Tech. Sci. 8–1(89), 44–49 (2021)
18. Lekanova, T.L., Andronov, A.V.: Practical recommendations on the use of wood waste for electricity production. Fundam. Res. 10-2, 239–244 (2017)
19. Kol'nichenko, G.I., Tarlakov, Y.V., Sirotov, A.V.: Bioenergetic technologies and forest industry complex. Forest. Bull. **21**(5), 64–68 (2017). https://doi.org/10.18698/2542-1468-2017-5-64-68
20. Haq, Q.A.U.: Design and implementation of solar tracker to defeat energy crisis in Pakistan. Int. J. Eng. Manuf. **9**(2), 31–42 (2019). https://doi.org/10.5815/ijem.2019.02.03
21. Subhonberdiev, A.S., Titova, E.V., Egorova, G.N.: Trends and prospects for the development of renewable energy sources in the European Union. Vestnik VGUIT [Proc. VSUET] **82**(4), 366–370 (2020).https://doi.org/10.20914/2310-1202-2020-4-366-370.(in Russian)
22. Saxena, S., Khan, M.Z., Singh, R.: Green computing: an era of energy saving computing of cloud resources. Int. J. Math. Sci. Comput. **7**(2), 42–48 (2021). https://doi.org/10.5815/ijmsc.2021.02.05
23. Alakbarov, R.G.: Method for effective use of cloudlet network resources. Int. J. Comput. Netw. Inf. Secur. (IJCNIS) **12**(5), 46–55 (2020)
24. Rezaeipanah, A., Nazari, H., Abdollahi, M.: Reducing energy consumption in wireless sensor networks using a routing protocol based on multi-level clustering and genetic algorithm. Int. J. Wirel. Microwave Technol. (IJWMT) **10**(3), 1–16 (2020)
25. Saxena, S., Khan, M.Z., Singh, R.: Green computing: an era of energy saving computing of cloud resources. Int. J. Math. Sci. Comput. (IJMSC) **7**(2), 42–48 (2021)
26. Rastorguev, I.P., Neizhmak, A.N.: Methodology of evaluation of the climatic potential of solar and wind energy. Heliogeophys. Res. **9**, 150–160 (2014)
27. Gavrishchenko, P.: Assessment of the wind energy potential of the Altai region. Works Young Sci. Altai State Univ. **13**, 37–40 (2016)
28. Shuklina, A.P., Shuklina, V.P., Ermolenko, B.V.: Development of a methodology for evaluating the fuel, electric power and heat power potentials of solid biofuel from plant and wood waste. Adv. Chem. Chem. Technol. **34**(11), 49–52 (2020)
29. Samarina, V., Skufina, T., Samarin, A., Tereshina, M.: Concerning experience and prospects of development of biogas power engineering in one of Russia's energy-deficiency agricultural regions. In: Conference: 19th International Scientific Conference Engineering for Rural Development (2020). https://doi.org/10.22616/ERDev.2020.19.TF015

A Concept of Cloud Knowledge Portal for Intelligent Decision Support in Additive Lattice Structure Formation from Aluminum Powder

Yuriy N. Kulchin, Valeria V. Gribova, Vadim A. Timchenko(✉), Marina V. Polonik, Dmitry S. Pivovarov, Dmitry S. Yatsko, Pavel A. Nikiforov, and Alexander I. Nikitin

Institute of Automation and Control Processes Far-Eastern Branch of the Russian Academy of Sciences (IACP FEB RAS), 5, Radio Street, Vladivostok 690041, Russia
vadim@dvo.ru

Abstract. The paper presents the research results of laser-based additive process Directed Energy Deposition (DED) aimed at improving the mechanical properties of thin-walled parts made of sheet AMg3 aluminum alloy, by creating on the surface of the lattice structure of powder material based on aluminum AK4-1. Taking into account the influence of a number of key technological parameters for laser powder-based additive processes of DED type on the properties of the final product and the lack of a proper level of intelligent support for such processes, the concept of cloud-based knowledge portal for intelligent support of decision-making by the equipment operator has been proposed. The concept is based on a two-level ontological approach to the formation of data- and knowledge bases of laser powder-based DED additive manufacturing. A brief description of the information bases and tools for the decision support in laser-based additive manufacturing is given. Mathematical modeling of the distribution process of thermal fields formed by interaction of laser radiation with the processed material is performed. Software that implements numerical calculations for the presented mathematical model of the DED process was created. A remote interaction scheme of intelligent decision support service for laser-based additive manufacturing processes with the created software has been proposed. The obtained results of numerical calculations in the Wolfram Mathematica technical computing system together with the expert information that is accumulated in the knowledge portal of the IACPaaS cloud platform are promising for training and decision-making by users of laser technological equipment.

Keywords: Laser technology · Additive manufacturing · Directed energy deposition · Decision support system · Laser powder cladding

1 Introduction

Modern trends in the development of machine-building manufacturing technologies are aimed at solving the problems of designing and creating metal parts with high mechanical properties (strength, stiffness etc.) and with minimal weight. The existing technological

Z. Hu et al. (Eds.): AIPE 2021, LNDECT 119, pp. 207–218, 2022.
https://doi.org/10.1007/978-3-030-97064-2_20

processes of traditional (subtractive) part manufacturing by casting, forging, stamping, and machining limit the opportunities for engineers and designers in creating lightweight products with complex geometry and multifunctional design capabilities. Fast-developing laser additive manufacturing (AM) technologies provide more flexibility for manufacturing products with complex structure, allow to minimize energy consumption, costs and amount of waste, and give engineers more freedom for manufacturing new lattice/cellular structures with complex geometry [1, 2].

The problem of increasing the stiffness and strength of thin-walled hull components without a significant increase in their mass can be solved by reinforcing the surface. The quality of three-dimensional metal products created by AM Directed Energy Deposition (DED) technology (ASTM 52900:2015, Additive Manufacturing – General Principles – Terminology) using high-power laser radiation is formed in two stages.

The articles published in the special issue "Modeling and Simulation for Additive Manufacturing" of Additive Manufacturing journal present plenty of research findings on mathematical modeling of AM processes, but "the role of modeling and simulation will remain critical to the development of AM technology for the foreseeable future" [3].

The first stage is characterized by making the correct technical decision by the operator of the technological complex when choosing the operation modes of laser, robotic and auxiliary equipment taking into account the thermophysical characteristics of the processed metal. The problem of creating a physical object according to its digital 3D model via layer-by-layer adding of powder material consists in the fact that the operator at the first stage performs a mathematical calculation of temperature field distribution on the surface and on the depth of the clad material. While using this approach, the obtained values of key equipment parameters should not include cases of intensive volumetric boiling and evaporation, which might lead to ejecting metal from the melt zone and forming structural defects in the deposited layer.

The second stage of creating quality bulk metal products using DED-technology, satisfying the requirements of regulatory and design documentation, is characterized by the problem of maintaining the values of design parameters throughout the duration of the additive process.

Considering the complexity of the decision-making process by an AM system user on the setting of key equipment parameters and the lack of a proper level of intelligent support for such a process [4], the team of authors are developing models, methods and cloud infrastructure for intelligent support of AM of metal parts from metal-powder compositions using laser technological equipment.

The paper presents the research results of the technical decision-making process by the laser-based additive manufacturing system (LBAMS) operator when forming a periodic lattice structure on the surface of a thin-walled aluminum alloy plate. To provide intelligent support for such kind of processes, a general concept and architecture of the cloud-based technological knowledge portal for the LBAM using laser powder-based DED (LPDED) technology has been proposed.

2 Material and Methods

The creation of a periodic lattice structure with specified cell sizes on the surface of a thin-walled plate of AMg3 aluminum alloy was performed using a LBAM system and DED technology. Implementation of additive processes was performed under the control of laser and robotic equipment software.

2.1 Equipment, Materials and Software

The main equipment in LBAMS that implements the DED process, in which energy from an external laser radiation source is used to join metal powder materials by fusing them in the process of deposition, is a high-power ytterbium fiber laser and an industrial robot-manipulator. Laser equipment: high power CW fiber laser LS-1-K, wavelength 1.07 μ; IPGP FLW-D50 optical head with PRECITEC 4W powder feed module 4W "four-stream nozzle". The programmed movement of the optical head was carried out with a KUKA KR30HA 6-axis industrial robot. The powder feeder PF-7103 manufactured by Plazmoavtomatika was used to feed metal powder material into the fusion zone. The industrial robot and the fiber laser were controlled using KUKA ExpertTech 3.2, KUKA LaserTech 3.0 and LaserNet software.

A series of numerical calculations of temperature fields and boundaries of the melt pool was performed using a technical computing software system Wolfram Mathematica 11.3. Digital model of a single lattice structure line is made using SolidWorks software (SolidWorks Campus 500). Microstructure investigation of deposited material has been performed using the following equipment: IM4000 ion etching unit; scanning electron microscope S-3400N; PMT-3M microhardness tester (microhardness was determined by the Vickers method, with a load of 0.98 N).

One of the LBAM key problems is the choosing (correct setting) of technological modes. Taking into account the uniqueness of the products, "single-part production", the intelligent support for the LBAMS operators to reduce the number of errors, help them in choosing technological modes is an urgent task. Its implementation is possible with the use of artificial intelligence methods [5–8].

The authors propose a hybrid approach to intelligent support of the operators based on the use of knowledge engineering methods, case-based search by analogy, and inductive data generalization [9]. This approach is conceptually similar to the approach and methodology proposed by H. Ko, et al. [5]. However, the uniqueness of the concept and architecture of decision support software for LPDED AM processes proposed by the authors is that they are based on a two-level ontological technique to the development of data- and knowledge bases about the LBAM. Ontologies are clearly separated from databases and knowledge bases. First of all, cognitive engineers together with experts in LBAM form the necessary ontologies. Further, in terms of these ontologies, LBAM specialists can develop databases and knowledge bases with no middlemen (e.g. knowledge engineers, software developers) involved in the form that they understand (all information resources are represented by concept digraphs [10, 11]).

In accordance with the concept and architecture, a technological knowledge portal for LBAM has been developed (Fig. 1).

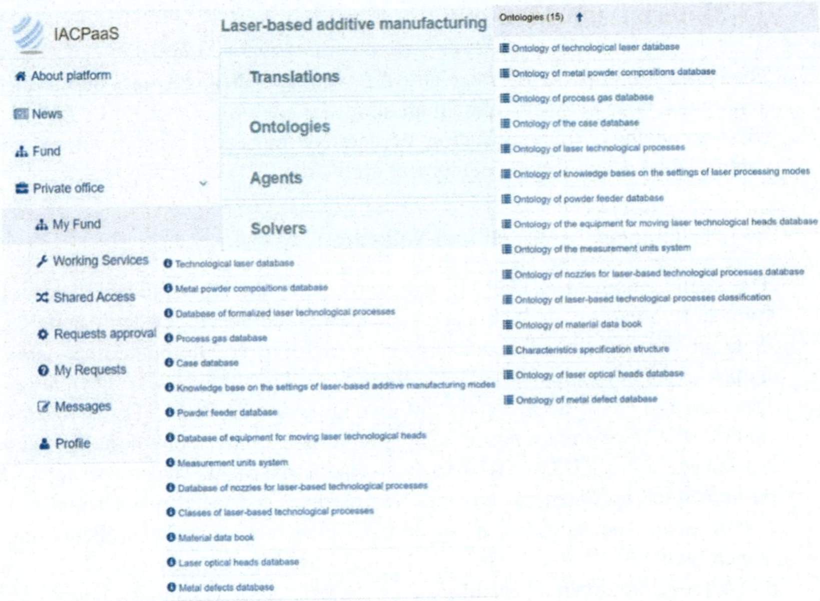

Fig. 1. Knowledge portal for laser-based additive manufacturing (the interface of the IACPaaS cloud platform website)

The portal is created on the IACPaaS (Intelligent Applications, Control and Platform as a Service) cloud platform [12] (https://iacpaas.dvo.ru) and includes information and software components. The information components of the knowledge portal are a set of databases and reference books used in LBAM, a case database, a knowledge base on the settings of LBAM modes, as well as a set of ontologies required for their formation.

Databases contain information about the components of a LBAMS (technological high-power lasers, laser optical heads, equipment for moving laser optical heads, e.g. industrial robots, powder feeders, etc.), and about the processed (consumable) materials for AM (metal powder compositions and process – carrier, shielding, compressing – gases).

The *knowledge base* contains formally represented relationships between (1) composition and properties of the processed parts, (2) composition and properties of the deposited powder materials, (3) gas environment that should be provided during deposition process, (4) deposition mode (a set of key process parameters), and the properties of the final products (parts). These kinds of dependencies have the form of production rules [13].

The *case database* contains a set of formalized protocols of technological operations, structured by groups depending on the success of their results.

The software components include editors for creating and maintaining databases and the knowledge base (driven by the corresponding ontologies), as well as a decision support system (DSS) based on both knowledge base and case database.

To date, a number of information and software components of the knowledge portal have been developed. Work on creating new components of the portal and improving the existing ones is ongoing. Thus, a set of ontology models has been developed for a unified and standardized (including terminological consistency) description of the following data and knowledge bases: databases of the equipment and processed (consumable) materials for a LBAM, the database of formalized protocols of technological operations (cases), the knowledge base on the settings of LBAM modes.

For the formal representation of ontologies, the *ontology description language* of the IACPaaS cloud platform is used [14, 15]. The tool "*Ontology editor*" is used for creating these ontologies in the platform's Fund (structured information repository) (Fig. 2).

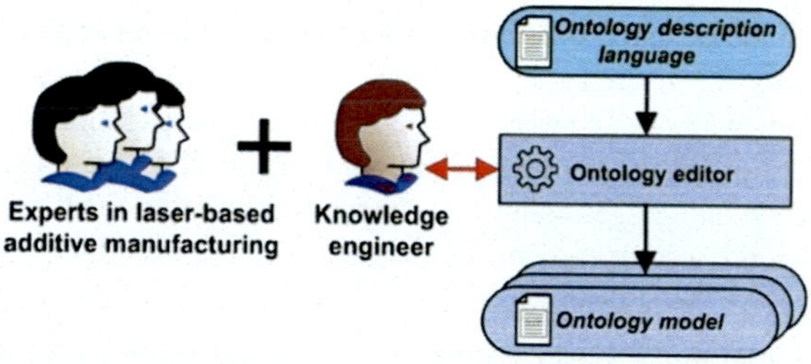

Fig. 2. The scheme of using the IACPaaS platform tool "*Ontology editor*" for creating the collection of ontologies in the platform's Fund

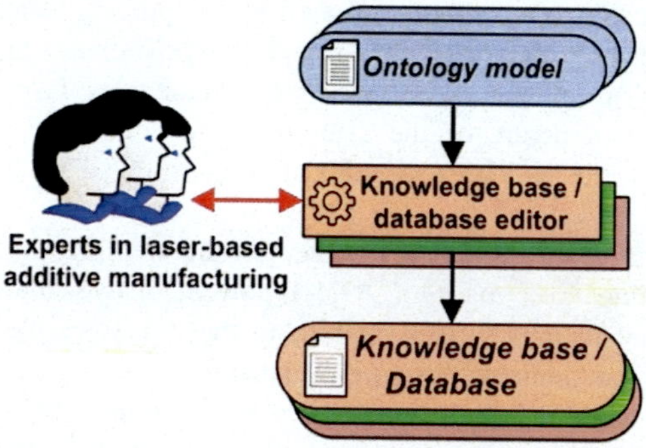

Fig. 3. Ontology-driven editors for databases and knowledge base

Editors for the formation and maintenance of databases, the knowledge base and the case database controlled by the corresponding ontologies have been developed (Fig. 3). Note that when changes are made to ontologies, the editors (their user interface and editing process) automatically adapt to these changes. With the use of editors, all databases and knowledge base are currently being formed.

2.2 Materials for Additive Manufacturing

The periodic lattice structure on the surface of a thin plate of AMg3 alloy was created from aluminum-based powder metal AK4-1. Powder from AK4-1 alloy is produced at the company "SphereM" (https://sferam74.ru) by spraying liquid metal. The particle size of the main fraction is 100–120 μ. Content of the main fraction is more than 90% by mass, density – 2.80 kg/dm^3. Given that aluminum is very sensitive to oxidation while being melted, the inert gas helium (He), which has good thermal conductivity and a high plasma generation threshold, was used as a shielding and carrier gas.

3 Design and Calculation of Laser Additive Process Parameters

The design of the lattice structure on the surface of a thin-walled plate was performed in two stages. At the first stage, the calculation of energy parameters for the creation of a single lattice line was performed. For this purpose, mathematical models reflecting the physical processes of the thermal fields distribution were used [16].

At the second stage, using the obtained calculation results, a single line of metal powder AK4-1 was formed on the surface of a 2.0 mm thick plate.

3.1 Calculation of Laser Additive Process Parameters for a Single Line

Based on the thermo-physical characteristics of the materials (Table 1) and laser treatment parameters (Table 2), determined by experts, calculation of additive building of a single line on the surface of a plate made of AMg3 aluminum alloy with thickness of 2.0 mm was performed.

When studying the influence of the LPDED process parameters on the structure of the formed lattice material it was found that preliminary heating of the substrate to temperature $T_0 = 100 \pm 20\ °C$ allows to obtain more favorable, in particular, more homogeneous, structure of the clad metal. In addition, this technological solution makes it possible to reduce the tendency of the clad material to crack and reduce porosity.

To determine the parameters of the temperature field, the melt pool and the height of the cladding layer made of AK4-1 powder, a two-step modeling algorithm and numerical calculations were proposed. In the first stage, the boundaries of the temperature field of the melt pool are calculated for an AMg3 substrate. The heating of the substrate by laser radiation is described by the differential heat conduction equation [17]. In Table 2 the preliminary parameters of laser cladding are determined by expert way (on the basis of thermo-physical characteristics of materials of the processed part and powder composition).

Our measured surface absorption coefficient of AMg3 aluminum alloy at a wavelength of 1.0 μm, at room temperature was 0.25 [18]. Taking into account that the initial temperature of the laser treatment process is 100 $°C$, on the basis of expert

Table 1 Thermal and physical characteristics of the sample of AMg3 aluminum alloy and powder alloy AK4-1 at T_0

	Thermal physical characteristics	Designation	AMg3 parameter values	AK4–1 parameter values	Units of measure
1	Material density	*p*	2670	2770	kg/m^3
2	Heat capacity	*c*	880	797	J/(kg·°C)
3	Thermal conductivity	*k*	151	146	J/(kg·°C)
4	Thermal diffusivity	$a = k/(cp)$	$6.427 \cdot 10^{-5}$	$6.613 \cdot 10^{-5}$	m^2/s
5	The size of the line to be clad				
	Height	*h*	0.002	0.000143·16 *layers*	m
	Length	*d*	0.1	0.05	m
	Width	*b*	0.07	0.0009	m
6	Initial temperature	T_0	100	100	°C
7	Melting temperature	T_p	639	645	°C
8	Boiling Point of metal	T_κ	2518	2518	°C
9	Coefficient surface absorption by AMg3 (1530) aluminum alloy at a wavelength of 1.0 μm	*A*	0.45		
10	Surface absorption coefficient of laser radiation by aluminum alloy AK4–1	*A*		0.45	

Table 2 Laser processing parameter values

	Laser processing parameters	Designation	Parameter values	Units of measure
1	Laser spot radius	r_0	0.00045	*m*
2	Laser power	P_0	650	*W*
3	Travel speed of the laser beam on the surface	*V*	0.012	*m/s*

evaluation, the surface absorption coefficient of laser radiation in AMg3 aluminum alloy in this experiment is taken equal to 0.45.

A series of numerical calculations in Wolfram Mathematica 11.3 was performed for thermal physical properties of the used materials and laser treatment parameters (Table 1 and Table 2). The melt pool depth is formed in a thin sample layer of $13.5 \cdot 10^{-5}$ m, while the material surface temperature must not exceed the boiling point T_b. The width of the melt pool on the plate surface is $\sim 0.00045 \cdot 2 = 0.0009$ m.

The second step solved the problem of adding a layer of powder to the workpiece in the formed melt pool. The new deposited layer creates a new object on the previous one, which is bounded by the intersection of the powder flow and the melt pool.

It is assumed that the added layer coincides with the temperature under the layer. We will consider that in one pass of the laser beam the thickness of the formed deposited layer is small enough ~0.143 mm (Table 1), and the layer material is uniformly distributed over the area of the melt pool. It was also assumed that the powder enters the melt pool already in the molten state, which allowed us to use the thermo-physical characteristics of powder material of AK4-1 alloy in calculations.

The calculated depth of the melt pool is $13.3 \cdot 10^{-5}$ m. The calculated temperature at the surface of the material was 1880 °*C*, i.e. with the increase of the cladding material there is an increase in the temperature of the molten metal. If the value of the melt temperature exceeds the value of $T_b = 2518$ °C (Table 1), there will be intensive volumetric boiling and evaporation, leading to metal ejection from the melt zone and the appearance of structural defects of the deposited layer.

3.2 Performing Single Line Laser Cladding

The calculations have shown that at the stage of determining the parameters of manufacturing one lattice line, it is advisable to reduce the laser power (Table 2). For the manufacture of one line on the surface of a thin-walled plate made of aluminum alloy, we will take the following parameters of the AM system. Laser radiation power is 630 W. Gas consumption is 30 l/min. Metal powder consumption is 2.2 g/min. Travel speed of laser beam on the surface – 0.012 m/s. Step of displacement along Z-axis – 0.13 mm. Initial temperature of the substrate is 100 ± 20 °*C*.

After the cladding of one line, its quality is evaluated by examining the microstructure. The analysis of the sample microstructure showed that the material obtained by laser powder cladding from AK4-1 powder has a much more dispersed structure than the alloy obtained by standard technology. The Vickers microhardness of the sample obtained by laser powder cladding was also measured. It was also compared with the microhardness of a sample made of aluminum wrought alloy AK4-1. Considering that the alloy produced by the additive LPDED process was not heat-treated, hardened, or artificially aged, the measured microhardness value (mean 103,6 ± 2,30 vs 129,3 ± 6,30) can be considered acceptable.

4 Construction of a Lattice Periodic Structure on the Surface of an Aluminum Alloy Plate

Based on the results of calculation of the parameters of the LPDED process and the results of the experiment on creating a separate line with the calculated parameters, taking into account the positive assessment of the microstructure and microhardness of the material obtained, a periodic lattice structure of the AK4-1 metal powder was created on the surface of a thin-walled aluminum (AMg3) plate (Fig. 4).

Additional information to the lattice structure process: number of deposited layers in a single line – 16; diameter of the laser spot on the processed surface – 0.9 ± 0.1 mm; length of one clad line – 60.0 ± 3.0 mm; distance between the lines – 11.0 ± 0.2 mm;

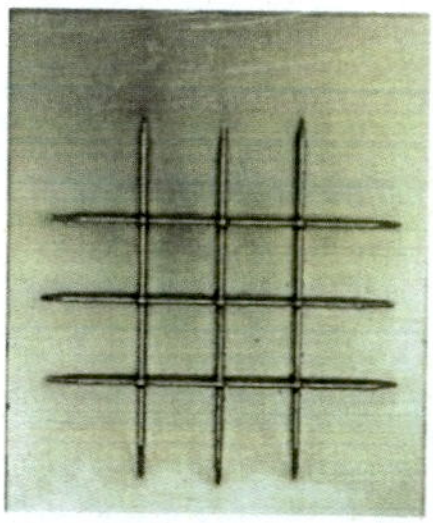

Fig. 4. Periodic lattice structure of the AK4-1 metal powder on the surface of a thin-walled plate of aluminum alloy

5 Discussion of the Research Results

The presented methodology for designing the technological process of laser 3D printing demonstrates an approach to the problem of setting the values of key equipment parameters when creating a volumetric periodic structure from metal powder on the surface of a thin-walled metal.

At the first stage, taking into account the thermophysical properties of the materials used and the equipment characteristics, the operator calculates the parameters of the LBAMS. Based on the results of a series of numerical calculations of temperature fields and boundaries of the melt pool, the operator decides whether the energy parameters of the external laser source are correct according to the criterion of the current value of the melt temperature: $T_k > T(r,z,t) > T_p$. In our work we used a technical computing software system Wolfram Mathematica 11.3.

To study the possibility of interaction of the DSS (included in the knowledge portal) with software tools for modeling and numerical computations of physico-chemical processes (accompanying the LPDED technology), a scheme for using software algorithm for calculating the parameters of the temperature field, of the melt pool and the height of the deposited layer from AK4-1 powder is proposed. The scheme is based on the technology for the interaction of IACPaaS platform services with external (in relation to the cloud platform) software [19]. The software implementation of the algorithm (external software) is a script in the Wolfram Language of the Wolfram Mathematica 11.3 system (Fig. 5).

A *middleware* is bundle of a *web-server* program, some *FastCGI gateway* for certain programming language and an application on this language. Such an application (a *software wrapper*) implements the launch of the *Wolfram Language script* with the transfer of data to it – the thermo-physical characteristics of the materials used and the laser processing parameters (see Table 1 and Table 2) received from the DSS via a web-server, and the return of the script results to the web-server (which will redirect it back to the DSS – IACPaaS platform service).

To implement this mechanism, we chose a software tool (web-server) that implements the processing of HTTP requests (Apache HTTP Server), and a FastCGI gateway, in the programming language of which the software wrapper is developed. Then they are installed and integrated on the node on which the Wolfram Mathematica 11.3

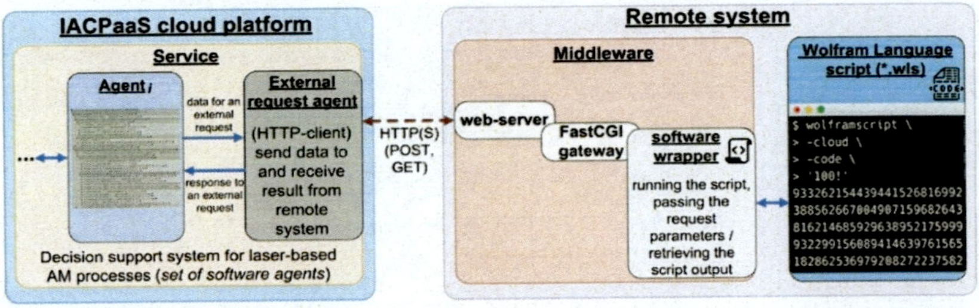

Fig. 5. A remote interaction scheme of the IACPaaS cloud platform service (DSS) with the Wolfram Language script

system is installed and which is accessible over Internet. The PHP programming language is used as FastCGI implementation because of its high-level methods for processing HTTP requests and running executable application files.

The second stage is the verification of the accepted parameters of the LBAMS. After manufacturing, a fragment of a part with a given geometric shape, the required properties of the obtained alloy are investigated. If the basic geometric shape and properties of the obtained material meet the specified requirements of the terms of reference, the operator proceeds to the next stage, in our case, to the creation of a three-dimensional periodic structure from metal powder on the surface of a thin-walled metal.

The third stage is the creation of a physical spatial product by successive addition of metal powder material. In practice, control of the reliability of the main working properties of the part obtained in the LPDED process is carried out with the use of nondestructive testing means.

The results of the study, in particular the formalized protocol of technological operation for the formation of a lattice periodic structure on the surface of an AMg3 aluminum alloy, are accumulated in the corresponding databases of the knowledge portal and serve to replenish the case database. The protocol is formed by the LBAMS operator after the completion of the technological operation. The formation of protocols is performed with the use of the editor controlled by the ontology of technological operations formalized protocols.

The case database will be used for case-based reasoning (search by analogy) and for forming the knowledge base using inductive data generalization methods (when the case database size becomes sufficient for training). So, the permanent accumulation of information in the shared technological space is an important task.

6 Summary and Conclusion

The presented methodology for laser additive design process of lattice periodic structure formation on the surface of an aluminum alloy plate is a part of research aimed at studying the methods of combination of matrix and reinforcing elements formed in the zone of intense interaction of laser radiation with composite system.

The knowledge portal on the IACPaaS cloud platform accumulates all the necessary information about technological processes (operations) of LBAM in a shared technological space. This makes it easier for the entire interested community to access this information [20]. The set of tools and technologies (toolkit) provided by the platform, the use of the ontological approach allow domain specialists to form and improve databases and the knowledge base in understandable conceptual structure and terminology, as well as provide intelligent support to LBAMS operators.

Data and knowledge bases of the portal will be useful in the process of training laser equipment users, and formalized representation of knowledge and data will provide the possibility of using this information by decision support software systems.

The results of this work are planned to be used for efficient operation of expensive high-tech equipment and raw materials, reducing the number of errors, increasing productivity and the level of automation of engineers-technologists and operators of laser technological equipment work.

Acknowledgment. The work was partially supported by Russian Foundation for Basic Research, project no. 20-01-00449.

References

1. Gibson, I., Rosen, D., Stucker, B.: Additive Manufacturing Technologies. Springer, New York (2010). https://doi.org/10.1007/978-1-4419-1120-9
2. Gu, D.D., Meiners, W., Wissenbach, K., Poprawe, R.: Laser additive manufacturing of metallic components: materials, processes and mechanisms. Int. Mater. Rev. **57**(3), 133–164 (2012)
3. To, A.C.: Preface. Add. Manuf. **12**(Part B), 143 (2016). https://doi.org/10.1016/j.addma.2016.09.003. Accessed 13 Dec 2021
4. Yang, L., et al.: Additive Manufacturing of Metals: The Technology, Materials, Design and Production. Springer, Cham (2017). https://doi.org/10.1007/978-3-319-55128-9
5. Ko, H., Witherell, P., Lu, Y., Kim, S., Rosen, D.W.: Machine learning and knowledge graph based design rule construction for additive manufacturing. Add. Manuf. **37**, 101620 (2021)
6. Sanfilippo, E.M., Belkadi, F., Bernard, A.: Ontology-based knowledge representation for additive manufacturing. Comput. Ind. **109**, 182–194 (2019)
7. Wang, Y., Zheng, P., Peng, T., Yang, H., Zou, J.: Smart additive manufacturing: current artificial intelligence-enabled methods and future perspectives. Sci. China Technol. Sci. **63** (9), 1600–1611 (2020). https://doi.org/10.1007/s11431-020-1581-2
8. Qi, X., Chen, G., Li, Y., Cheng, X., Li, C.: Applying neural-network-based machine learning to additive manufacturing: current applications, challenges, and future perspectives. Engineering **5**, 721–729 (2019)
9. Gribova, V., Kulchin, Y., Nikitin, A., Timchenko, V.: The concept of support for laser-based additive manufacturing on the basis of artificial intelligence methods. In: Kuznetsov, S.O., Panov, A.I., Yakovlev, K.S. (eds.) Artificial Intelligence, vol. 12412, pp. 403–415. Springer, Cham (2020). https://doi.org/10.1007/978-3-030-59535-7_30
10. Rajangam, E., Annamalai, C.: Graph models for knowledge representation and reasoning for contemporary and emerging needs – a survey. Int. J. Inf. Technol. Comput. Sci. **8**(2), 14–22 (2016)

11. Hepsiba Mabel, V., Selwyn, J.: A review on the knowledge representation models and its implications. Int. J. Inf. Technol. Comput. Sci. **8**(10), 72–81 (2016)
12. Diaby, T., Rad, B.B.: Cloud computing: a review of the concepts and deployment models. Int. J. Inf. Technol. Comput. Sci. **9**(6), 50–58 (2017)
13. Aydilek, I.B., Gümüşçü, A.: Designing a rule based expert systems for contact lenses patients. Int. J. Inf. Technol. Comput. Sci. **10**(3), 18–26 (2018)
14. Gribova, V.V., Kleshchev, A.S., Moskalenko, F.M., Timchenko, V.A.: A two-level model of information units with complex structure that correspond to the questioning metaphor. Autom. Doc. Math. Linguist. **49**(5), 172–181 (2015). https://doi.org/10.3103/S0005105515050052
15. Gribova, V.V., Kleshchev, A.S., Moskalenko, F.M., Timchenko, V.A.: A model for generation of directed graphs of information by the directed graph of metainformation for a two-level model of information units with a complex structure. Autom. Doc. Math. Linguist. **49**(6), 221–231 (2015). https://doi.org/10.3103/S0005105515060059
16. Yatsko, D.S., Polonik, M.V., Dudko, O.V.: Optimization of fiber laser parameters at local high-temperature exposure to metal. In: Proceedings of SPIE. 10176. Asian-Pacific Conference on Fundamental Problems of Opto- and Microelectronics (2016). 1017624
17. Malashchenko, A.A., Mezenov, A.V.: Laser Welding of Metals. Mashinostroenie, Moscow (1984).(in Russian)
18. Kulchin, Y.N., Nikitin, A.I., Subbotin, E.P.: Laser underwater cleaning of marine vessels hulls. Appl. Photonics **4**, 90–104 (2020). (in Russian)
19. Gribova, V.V., Moskalenko, F.M., Timchenko, V.A., Fedorishchev, L.A.: Technology of interaction of IACPaaS cloud platform services with external software. Softw. Syst. **31**(2), 233–238 (2018). (in Russian)
20. Bhatti, H.J., Rad, B.B.: Databases in cloud computing: a literature review. Int. J. Inf. Technol. Comput. Sci. (IJITCS) **9**(4), 9–17 (2017). https://doi.org/10.5815/ijitcs.2017.04.02

Author Index

Z. Hu et al. (Eds.): AIPE 2021, LNDECT 119, pp. 219–220, 2022.
https://doi.org/10.1007/978-3-030-97064-2